Topics in Applied Physics

Volume 136

Topics in Applied Physics is a well-established series of review books, each of which presents a comprehensive survey of a selected topic within the area of applied physics. Edited and written by leading research scientists in the field concerned, each volume contains review contributions covering the various aspects of the topic. Together these provide an overview of the state of the art in the respective field, extending from an introduction to the subject right up to the frontiers of contemporary research.

Topics in Applied Physics is addressed to all scientists at universities and in industry who wish to obtain an overview and to keep abreast of advances in applied physics. The series also provides easy but comprehensive access to the fields for newcomers starting research.

Contributions are specially commissioned. The Managing Editors are open to any suggestions for topics coming from the community of applied physicists no matter what the field and encourage prospective book editors to approach them with ideas.

2018 Impact Factor: **0.746**

More information about this series at http://www.springer.com/series/560

Kaoru Yamanouchi · Dimitrios Charalambidis
Editors

Progress in Ultrafast Intense Laser Science XV

Springer

Editors
Kaoru Yamanouchi
Department of Chemistry
School of Science, The University of Tok
Tokyo, China

Dimitrios Charalambidis
FORTH-IESL
Heraklion, Greece

ISSN 0303-4216 ISSN 1437-0859 (electronic)
Topics in Applied Physics
ISBN 978-3-030-47097-5 ISBN 978-3-030-47098-2 (eBook)
https://doi.org/10.1007/978-3-030-47098-2

This Springer imprint is published by the registered company Springer Nature Switzerland AG
The registered company address is: Gewerbestrasse 11, 6330 Cham, Switzerland

Preface

We are pleased to present the fifteenth volume of Progress in Ultrafast Intense Laser Science. As the frontiers of ultrafast intense laser science rapidly expand ever outward, there continues to be a growing demand for an introduction to this interdisciplinary research field that is at once widely accessible and capable of delivering cutting-edge developments. Our series aims to respond to this call by providing a compilation of concise review-style articles written by researchers at the forefront of this research field, so that researchers with different backgrounds as well as graduate students can easily grasp the essential aspects.

As in the previous volumes, each chapter of this book begins with an introductory part, in which a clear and concise overview of the topic and its significance is given, and moves onto a description of the authors' most recent research results. All chapters are peer-reviewed. The articles of this fifteenth volume cover a diverse range of the interdisciplinary research field, and the topics may be grouped into four categories: laser-induced filamentation (Chaps. 1 and 2); atoms and molecules in a laser field (Chaps. 3 and 4); interaction of solid materials with a coherent light field (Chaps. 5 and 6); and ion acceleration and ionization of atoms in super intense laser fields (Chaps. 7 and 8).

From the third volume, the PUILS series has been edited in liaison with the activities of the Center for Ultrafast Intense Laser Science at the University of Tokyo, which has also been responsible for sponsoring the series and making the regular publication of its volumes possible. From the fifth volume, the Consortium on Education and Research on Advanced Laser Science, the University of Tokyo, has joined this publication activity as one of the sponsoring programs. The series, designed to stimulate interdisciplinary discussion at the forefront of ultrafast intense laser science, has also collaborated since its inception with the annual symposium series of ISUILS (http://www.isuils.jp/), sponsored by JILS (Japan Intense Light Field Science Society).

We would like to take this opportunity to thank all of the authors who have kindly contributed to the PUILS series by describing their most recent work at the frontiers of ultrafast intense laser science. We also thank the reviewers who have read the submitted manuscripts carefully. One of the co-editors (KY) thanks Ms. Mihoshi Abe for her help with the editing processes.

We hope this volume will convey the excitement of ultrafast intense laser science to the readers, and stimulate interdisciplinary interactions among researchers, thus paving the way to explorations of new frontiers.

Tokyo, Japan — Kaoru Yamanouchi
Heraklion, Greece — Dimitrios Charalambidis

Contents

Contributors

Tim Bayer Institut für Physik, Carl von Ossietzky Universität Oldenburg, Oldenburg, Germany

S. V. Bulanov Institute of Physics of the ASCR, ELI-Beamlines project, Prague, Czech Republic

See Leang Chin Center for Optics, Photonics and Lasers (COPL), Laval University, Quebec City, Canada

M. F. Ciappina Institute of Physics of the ASCR, ELI-Beamlines project, Prague, Czech Republic

T. Ditmire Department of Physics, University of Texas at Austin, Center for High Energy Density Science, Austin, TX, USA

Kevin Eickhoff Institut für Physik, Carl von Ossietzky Universität Oldenburg, Oldenburg, Germany

Yuji Fukuda Kansai Photon Science Institute, National Institutes for Quantum and Radiological Science and Technology, Kizugawa-shi, Kyoto, Japan

Attila G. Császár Institute of Chemistry, Eötvös Loránd University and MTA-ELTE Complex Chemical Systems Research Group, Budapest, Hungary

Kunio Ishida School of Engineering, Utsunomiya University, Utsunomiya, Tochigi, Japan

Gábor J. Halász Department of Information Technology, University of Debrecen, Debrecen, Hungary

Masato Kanasaki Graduate School of Maritime Sciences, Kobe University, Kobe, Japan

Stefanie Kerbstadt Institut für Physik, Carl von Ossietzky Universität Oldenburg, Oldenburg, Germany

G. Korn Institute of Physics of the ASCR, ELI-Beamlines project, Prague, Czech Republic

Erik Lötstedt Department of Chemistry, School of Science, The University of Tokyo, Tokyo, Japan

Keiji Oda Graduate School of Maritime Sciences, Kobe University, Kobe, Japan

Tamás Szidarovszky Institute of Chemistry, Eötvös Loránd University and MTA-ELTE Complex Chemical Systems Research Group, Budapest, Hungary

Parinda Vasa Department of Physics, Indian Institute of Technology Bombay, Mumbai, India

Ágnes Vibók Department of Theoretical Physics, University of Debrecen, Debrecen, Hungary;
ELI-ALPS, ELI-HU Non-Profit Ltd., Szeged, Hungary

S. Weber Institute of Physics of the ASCR, ELI-Beamlines project, Prague, Czech Republic;
School of Science, Xi'an Jiaotong University, Xi'an, China

Matthias Wollenhaupt Institut für Physik, Carl von Ossietzky Universität Oldenburg, Oldenburg, Germany

Kaoru Yamanouchi Department of Chemistry, School of Science, The University of Tokyo, Tokyo, Japan

Tomoya Yamauchi Graduate School of Maritime Sciences, Kobe University, Kobe, Japan

Youyuan Zhang Department of Chemistry, School of Science, The University of Tokyo, Tokyo, Japan

Chapter 1
Femtosecond Laser Filamentation Induced Phenomena and Applications

See Leang Chin

Abstract The temporal evolution of a filament is followed. The physical processes evolve from quantum effect in the fs time scale through chemistry in the ps and ns time scale to classical physics in the microsecond to second time scale. A few applications and phenomena arising from each step of the physical evolution are discussed briefly.

1.1 Introduction

Since the accidental discovery, in 1995, of long distance self-focusing of a femtosecond (fs) Ti-sapphire laser pulse in air [1] by the Michigan group led by Prof. Gerard Mourou, Nobel Laureate 2018, a new sub-field of nonlinear optics, namely, filamentation, was born [2 7]. Filamentation is a phenomenon of the propagation of a powerful fs laser pulse in a high density transparent medium such as atmospheric pressure air or condensed media. The most popular laser used in these works is the Ti-sapphire laser. We shall limit the discussion in the present work to the propagation of this laser in air. The wavelength of the Ti-sapphire laser is around 800 nm; hence, its single photon energy is around 1.55 eV. This photon energy is much smaller than the ionization potentials of the molecules in air (the ionization potential of oxygen molecule is 12.1 eV and that of nitrogen molecule, 15.6 eV). Self-focusing of the laser pulse would result in a self-focal region in which the intensity is so high that tunnel ionization of the O_2 and N_2 molecules occurs [8]. The negative index of refraction of the plasma will defocus the laser pulse and stop the self-focusing. HOKE (Nonlinear Higher Order Kerr Effect) would probably not play a competitive role before ionization because of the fast rise of the laser pulse to a high field regime. The latter corresponds to the rapid onset of the non-perturbative regime. That is to say, the perturbative description (hence, the HOKE effect) cannot be used anymore [9]. Many physical processes are involved from the birth to the

S. L. Chin (✉)
Center for Optics, Photonics and Lasers (COPL), Laval University, Quebec City, Canada
e-mail: slchin@phy.ulaval.ca

K. Yamanouchi and D. Charalambidis (eds.), *Progress in Ultrafast Intense Laser Science XV*, Topics in Applied Physics 136,
https://doi.org/10.1007/978-3-030-47098-2_1

end of the decay of a filament. In this paper, the author will follow the physical evolution of a filament from the beginning to the end while briefly pointing out a limited number of applications or phenomena along the way without going into any detail. It is not meant to be complete. Hopefully, this would stimulate new ideas for new physics and applications. Filamentation starts with 'instantaneous' quantum mechanical ionization/excitation/dissociation in the fs time scale followed by a fast decay in the picosecond and nanosecond time scale. Classical thermodynamic decay would then follow till the end when the medium comes back to the initial condition. It is assumed that the readers are familiar with the physical evolution of a filament excited by a fs Ti-sapphire laser pulse in atmospheric pressure air [2–7, 10].

1.2 High Intensity Short Pulse Interaction

Self-focusing of a powerful short laser pulse in air (transparent medium) would in principle keep on self-focusing into a geometrical singularity if there were no other physical mechanism that would defocus the pulse. If this were the case, we could imagine that the focal area would be very small. Furthermore, we knew that we could practically put in a lot of energy into a laser pulse by using laser amplifiers. This would make the intensity inside the self-focal zone extremely strong, so strong that we could imagine it would become a formidable 'almighty bullet' capable of piercing through any material at a distance. Since there seemed to be no theoretical limit in putting as much energy as possible into the laser pulse, even nuclear reaction could be envisioned on the target. Such was the naïve thought (dream) of the author at the beginning of the 'filamentation era'. He even 'generously' gave out this 'great original' idea to some applied scientists who were eager to get new exciting ideas, brought them home and executed them quickly. Unfortunately, life was more complicated. Not long after, the author realized that the intensity was clamped at the self-focal zone [11]. The clamped intensity in the filament core was limited to about 5×10^{13} W/cm^2 [11] for a free propagating pulse or a few times 10^{14} W/cm^2 if an external lens was used [12].

Although the author's dream was broken, the intensity inside the filament core was still very respectable because it could ionize even nitrogen molecules whose ionization potential was rather high (15.6 eV). In fact, it is because of such ionization of the oxygen and nitrogen molecules that gives rise to a plasma. This plasma, having a negative refractive index, would defocus the laser pulse or slow down the self-focusing. Further self-focusing would increase the intensity and would result in more ionization, hence, higher plasma density or stronger de-focusing. The plasma defocusing effect would eventually stop the self-focusing effect. At this balancing point, the spot size is minimum and the intensity maximum. Further self-focusing would increase the plasma density which would now defocus the pulse or enlarge the spot size resulting in a lower intensity. Because the laser pulse has a power distribution in the temporal domain or spatial distribution along the propagation direction, using the slice-by-slice self-focusing model [7], each slice of the front part of the pulse will self-focus at a different position along the propagation axis resulting in a continuous 'series' of self-focal spots. At each spot, the defocusing effect of the plasma balances the self-focusing effect. The consequence is a continuous series of self-focal spots that become a linear narrow nearly cylindrical zone, hence, a filament along the propagation direction. Inside the continuous series of spots or the filament core, the intensity is constant or clamped [11]. This clamped intensity is independent of pressure [13]. This is a surprising result. Note that only the lowest fundamental spatial mode of the laser undergoes filamentation [7, 14] while the pulse along the filament core is a continuous propagating series of plane waves [7].

Since the defocusing effect of a plasma is much stronger than the self-focusing effect, it doesn't take a high density plasma to balance the self-focusing effect. In fact, according to our experimental measurement [12], depending on the strength of external focusing, the plasma density is two to three orders of magnitude lower than the density of atmospheric pressure air; i.e. most of the molecules inside the filament are neutrals. These neutral air molecules 'bathe' in the high clamped intensity zone. This is a very special environment in which a large number of neutral molecules could interact with a high laser intensity for a short time. In this environment, each molecule would undergo uni-molecular interaction with the high laser intensity as if it were in a vacuum system because the laser pulse duration (fs time scale) is much shorter than the mean-free-time of an electron undergoing inverse Bremsstrahlung process (picosecond time scale) [7]. This would avoid laser induced breakdown. That is to say, the plasma inside the filament comes from uni-molecular interaction with the laser. Such an environment is unique and cannot be found elsewhere. We shall look at some consequences of such interactions. But before we do so, let us look at the temporal development of the filament core.

At the beginning, we can imagine the filament core along the filament axis as an imaginary cylindrical zone filled with neutrals. When the laser pulse arrives, the slice-by-slice self-focusing mechanism essentially means that each slice successively self-focuses into the imaginary cylinder. Each slice injects a flux of electromagnetic (EM) field from the side into the cylinder resulting in a high intensity zone in the form of a thin cylindrical slab. We idealize that the thin slab of high intensity EM flux fits nicely in the imaginary cylindrical zone. This slab is at least one wavelength thick and the diameter is the same as that of the filament zone (the imaginary cylindrical zone); the diameter is about 50–100 μm. Inside this slab, ionization takes place 'instantaneously' and the EM flux is then forced out (diverges out) of the zone by the self-generated plasma. Each successive slice will 'inject' a flux of EM field becoming a 'slab' with the same intensity into successive new positions in the cylinder in the forward propagation direction and diverges out. It is as if a dot of high intensity light 'bullet' (slab of light) propagates through the imaginary filament zone. The propagation speed of the slab of EM field is the speed of light in the neutral medium; i.e. c/n_0 where c is the speed of light in vacuum and n_0 is the linear index of refraction of neutral air [7]. This is because each slab represents the zone in which intensity clamping takes place. When intensity clamping occurs, the nonlinear index in the neutrals and the plasma index of refraction cancel each other [7]. It takes a finite time for the successive slabs to sweep through the filament zone.

The first 'instantaneous' interaction is between the strong EM flux and the air molecules inside the slab. In each slab of EM flux, some known physical events would occur simultaneously, most of them being uni-molecular interaction (UMI) while a few are collective interactions (CI) related to the index of refraction of the gas, hence, the gas density. The following are some examples [7]: tunnel ionization and dissociation of molecules (UMI), nonlinear pumping of molecular excited states (UMI), nonlinear excitation of rotational wave packets and molecular alignment (UMI), odd harmonics generation (CI), self-phase modulation in the neutrals and in the plasma (super-continuum generation or white light laser) (CI), etc. Many of these events are measured through measuring optical signals along the filament axis; i.e. the summation of the signals from successive slabs of light along the filament. This means that the signal intensity is high. This is one big advantage of having a filament. In what follows, we shall discuss a few of the above mentioned phenomena. More detailed discussion can be found in [2–7, 10].

1.3 Air Lasing

Some of the primary events could not be detected directly. One has to measure the post-interaction signature and infer back to the primary events. For example, in air, after tunnel ionization of N_2 and O_2 [8], normally, the experimentalist will have to wait till the electrons or ions are collected and measured. In a high vacuum system, the pressure is low. An experimentalist could take his/her time (from nanosecond to microsecond) to make measurement of the electrons or ions using time-of-flight mass spectrometers without having any fear of collisions among these new particles or without having to worry about the collective effect of a plasma. In atmospheric pressure air, counting electrons and ions is too difficult because the mean free time of collision is too short (ps for electrons in one atmospheric air). One has to rely upon the measurement of the fluorescence emitted from the system and infer it to ionization. Moreover, by making the measurement along the propagation axis, the fluorescence signal would be enhanced. Interestingly, in measuring the fluorescence from nitrogen molecules after strong interaction inside the filament, some unexpected additional-uni-molecular interaction was observed.

The unexpected phenomena are explained in the following. Tunnel ionization of the nitrogen molecules results in population inversion in N_2 and $N_2{}^+$; this phenomenon is called air lasing [15]. This so-called air lasing comes from the fact that amplified spontaneous emission along the filament length could be detected easily [15]. However, the mechanism of air lasing is rather involved [16, 17]. Population inversion in the N_2 neutral fluorescence comes from collision in the plasma. The interaction is the following [16].

$$N_2{}^+ + N_2 \rightarrow N_4{}^+$$
$$N_4{}^+ + e \rightarrow N_2(C^3\Pi_u) + N_2$$

The emission from the $N_2(C^3\Pi_u)$ state to the empty $N_2(C^3\Pi_g)$ state (hence, population inversion) gives rise to fluorescence at 337 nm, etc. with gain.

Population inversion between an excited state and the ground state of the N_2^+ ions had been intriguing in recent years. In principle, both the ground and excited states of N_2^+ were prepared through tunnel ionization of neutral N_2 from the ground state. Normal wisdom would predict that the population in the excited state should be lower than that in the ground state [18]. However, experiment showed that there was population inversion [15]. The myth was recently clarified [17]. The population inversion was realized by the ionization of N_2 into the ground state $X^2\Sigma_g^+$ and the excited state $B^2\Sigma_u^+$ of N_2^+ [18] together with the simultaneous coupling among these two states and the intermediate state $A^2\Pi_u$. The intermediate $A^2\Pi_u$ state was essential in the coupling; it served as a population reservoir in the process of transferring the population from the ground $X^2\Sigma_g^+$ state into the intermediate $A^2\Pi_u$ state thus depleting the population from the ground $X^2\Sigma_g^+$ state; population inversion was thus established between the $B^2\Sigma_u^+$ state and the ground $X^2\Sigma_g^+$ state.

1.4 Population Inversion in the Air Filament Containing Other Molecules

Lasing action (or population inversion) inside the filament in air does not pertain only to nitrogen. Oxygen also exhibited population inversion in the fs intense field [19, 20]. In fact, we have observed that dissociation of H_2O molecules in the fs Ti-sapphire laser filament in a humid air gave rise to OH radicals which were excited into a state of population inversion. Gain was observed in the u.v. fluorescence (around 308.9 nm) from OH [21]. Another observation was that inside a filament of the second harmonic (400 nm) of a fs Ti-sapphire laser in air in the laboratory with some humidity, after the primary interaction (ionization and dissociation), subsequent collision would induce more chemical reactions. We observed that there was gain in the fluorescence (336.0 nm) from NH from inside the filament [22]. The excited NH obviously came from both ionization and dissociation of H_2O and N_2 as well as subsequent collision events.

The above observations have not been analyzed thoroughly yet. However, one could speculate a little. For example, to create a free radical such as NH after ionization and dissociation of N_2 and H_2O, H and N or their ions would have to collide to form the final product NH. For two ground state atoms, the collision would have to be sufficiently strong; i.e. the relative collision speed would have to be high so that a sufficient amount of kinetic energy would be absorbed by the colliding particles to overcome the potential barrier between two ground state atoms. Presumably, there is not much kinetic energy transferred from the laser field to the dissociated atoms during tunnel ionization and dissociation of the molecules inside the filament core. One possible scenario is Coulomb explosion. However, the probability is too low. This is because Coulomb explosion requires an intensity of the order of 10^{15}

W/cm^2 while the clamped intensity inside the filament core in air is too low (5×10^{13} W/cm^2). Most of the photon energies being absorbed would go to the electrons for ionization. At room temperature, the average kinetic energy of the molecules is in the range of meV while the energy for molecular bonding is of the order of eV. Consequently, if the initial states of both N and H were in the ground state, the two atoms would not be able to react or bond up together. One of them would have to be in an excited state. This means that the final product molecule would also be in an excited state. If this excited state were a radiative state, fluorescence would occur. The lower state of the transition being not prepared during the formation process, the system would naturally be inverted. Now if the collision gave rise to NH^+ ions, the collision with electrons inside the filament would result in a neutral NH in the Rydberg states which would cascade decay to a more stable excited state. If this excited state were radiative, it would fluoresce and decay to the lower state which initially was empty, hence population inversion. It is thus imaginable that similar population inversion could be observed in many other similar interaction involving small molecules in a filament.

When more complex molecules such as hydrocarbon gases CH_4, C_2H_2 and C_2H_4 are mixed in an air filament, amplified spontaneous emission, hence, population inversion, was observed in the fluorescing fragment CH [23]. This was explained as being due to 'instantaneous' excitation of the molecules into the super-excited states [24, 25] followed by very fast decay (fs time scale [26, 27]) into many channels, one of them being dissociation into neutrals one of them being excited. The super-excited states are believed to be excited through population trapping in a strong laser field in our recent analysis [28–30]. Energetically speaking, the fluorescing neutral fragments were prepared in the excited states after the breakup of the 'hot' parent molecules. The lower state of the radiative transition should be empty; i.e. population inversion was prepared 'naturally'. Reference [28] gives a review of this phenomenon.

1.5 Intensity Clamping, Self-mode Cleaning and Self-phase Locking

Inside a filament, only the lowest order spatial mode survives during the filamentation process while the intensity is high and clamped [7, 11]. The propagating thin slabs of intense EM flux are a continuous series of propagating high intensity plane wave fronts. This peculiar environment would favor efficient generation of new EM waves. We shall consider two examples to illustrate the physical idea of the efficient process, namely, third harmonic generation (THG) and four wave mixing (4WM). (See [7] for detail.) From nonlinear optics, we know that both these phenomena depends on the nonlinear polarization of the material $P^{(3)}$ where

$$P^{(3)} \propto \chi^{(3)} EEE \tag{1.1}$$

where E is the total electric field at a position inside the filament and $\chi^{(3)}$ is the third order susceptibility of the medium. Here, we assume that the pump Ti-sapphire laser pulse is linearly polarized and the third harmonic (TH) or the second laser pulse for four wave mixing also linearly polarized and parallel to the polarization of the pump field. As such, we use scalar description.

An efficient third harmonic generation would require that once a TH signal was created at the beginning of the filament in the forward propagating direction, it would faithfully follow the pump without any walk-off; i.e. the speed of propagation of the pump and that of the TH would be the same or almost the same. Such a situation would give rise to a steady pile-up or in-phase accumulation of the TH signals along the whole filament so that at the exit of the filament, the energy/intensity of the TH would be very high. However, the TH's frequency being very different from that of the pump, their propagation speeds c/n (c; speed of light in vacuum, n, index of refraction in the medium) are very different normally; i.e. the indices of refraction of the pump (frequency ω) and the TH (frequency 3 ω), namely, $n(\omega)$ and $n(3\omega)$, are very different. This would result in a walk-off of the pump and TH signals. At the end of the filament, there would be no in-phase accumulation of the TH signals. Destructive interference would set in at many places along the way. The final signal would be weak.

What is surprising is that, inside a filament, the speeds of propagation of the TH and the pump are almost equal; i.e. their indices of refraction are almost equal [7, 31]. Hence, in-phase accumulation of the TH signal along the full filament length would give rise to a very strong TH signal; i.e. very high conversion efficiency as compared to normal non-filamentation cases.

The reason why the indices of refraction of the pump and the TH inside the filament are almost equal is because of the high clamped intensity inside the filament. We note that the index of refraction is defined as

$$n = \sqrt{\varepsilon/\varepsilon_0} = \sqrt{1+\chi} = \sqrt{1+P/E} \tag{1.2}$$

where ε is the dielectric constant of the medium and χ, the total susceptibility of the medium. P is the total scalar material polarization and E is the total electric field at the appropriate frequencies. One can see from (1.2) that the larger the ratio P/E is, the larger the index n and the slower the speed of propagation will be. In other words, the larger the nonlinear polarization $P^{(3)}$ is, the larger the index of refraction will be. In air, we assume that a weak TH wave is generated at the beginning. We shall look at the amplitudes of the TH and pump waves only. For the sake of clarity, the effect of the low density plasma is neglected. The fundamental and the TH are assumed to be propagating in a neutral air medium.

$$P = \varepsilon_0 \chi^{(1)} E + \varepsilon_0 \left(\chi^{(3)} EEE + \text{higher order terms}\right) \tag{1.3}$$

$$E = E_\omega + E_{3\omega} \tag{1.4}$$

where E_ω and $E_{3\omega}$ are the amplitudes of the pump and TH, respectively. The third harmonic term would be

$$P^{(3)} \propto EEE = (E_\omega + E_{3\omega})^3 \tag{1.5}$$

After expansion of the last term in (1.5), there will be a term which is proportional to the clamped intensity E_ω^2 inside the filament. We shall call this term 'clamped' polarization;

$$\text{i.e.} \quad P^{(3)}_{\text{clamped}} \propto E_\omega^2 E_{3\omega} \tag{1.6}$$

This term would be the dominant term among the other terms of $P^{(3)}$ for TH generation because of the high clamped intensity E_ω^2. It would increase the index of refraction of the TH according to (1.2), hence, reducing the speed of propagation. Similarly and simultaneously, according to detailed calculation and analysis [7, 31], the index of refraction of the fundamental is also controlled by the clamped intensity and is increased. The consequence is that the two indices of refraction are almost equal; i.e. the nonlinear speed of propagation of the TH and that of the fundamental are almost equal. Since at the position where intensity clamping takes place, the fundamental is a plane wave [7], the TH is also a plane wave. Further analysis [7] shows that these two plane waves do not overlap. The TH plane wave front faithfully lags behind the fundamental plane wave in space by half the TH's wavelength. There is no walk-off. (Full calculation including the effect of the weak plasma showed that the indices of refraction of the fundamental and the TH are almost equal [7, 31].)

This qualitative analysis explains why the efficiency of TH generation inside a filament would increase by 2 orders of magnitude as compared to the case without filament [7, 31]. High clamped intensity is the key reason. We call this self-group phase locking which is the consequence of intensity clamping. In fact, we found that the TH was well stabilized in the sense that the root-mean-square fluctuation of the TH was much lower than that of the pump laser because of intensity clamping. We also called this self-stabilization [31].

Since TH generation is a special case of parametric processes involving $\chi^{(3)}$, we can now predict that any other parametric process in air (or in any other system with inversion symmetry) would give rise to self-group phase locking. An example is mixing a fs frequency tunable laser pulse in the i.r. with the pump pulse inside a filament. It is a four-wave mixing (4WM) process [7, 32]:

$$\omega_{4\text{WM}} = 2\omega_{800} - \omega_{\text{i.r.}} \tag{1.7}$$

where $\omega_{4\text{WM}}$ is the resultant frequency after mixing, ω_{800}, the pump Ti-sapphire laser frequency and $\omega_{\text{i.r.}}$, the tunable i.r. laser frequency. The tuning range of the i.r. laser spans from 1100 to 2400 nm. Experiment [32] showed that the 4WM pulse was tunable across almost the entire spectral range of the visible light. We define the conversion efficiency of the 4WM pulse as the ratio of the energy of the 4WM pulse over that of the i.r. pulse. It was found that the conversion efficiency without filament is less than 10^{-5} while that with filament was 30% in air. The reason is again 'filament guiding' (phase locking) of the 4WM pulse by the clamped intensity of the filament similar to the situation in TH generation.

Another advantage about filament guiding is that the spatial mode of the 4WM pulse has an M^2 almost equal to unity according to the experiment in [32]. This is because of self-mode cleaning inside the filament [14]. Only the lowest order mode propagates in the filament. Its coupling with the i.r. pulse would naturally produce a guided 4WM pulse through this single spatial mode; hence, the 4WM pulse necessarily would become a single lowest order mode pulse with $M^2 = 1$. For more detail, see [7, 14, 32].

We could thus predict that all parametric processes taking place in a filament would give rise to a high conversion efficiency with an almost perfect spatial mode.

1.6 Filament Induced Chemical Reaction

The strong laser field inside a filament would not only ionize but also fragment (dissociate) a molecule. In particular, if polyatomic molecules such as water molecules, hydrocarbons, etc. are mixed inside the ambient air, instantaneous ionization and dissociation of these molecules in the filament would be followed by slower processes of collisions. New chemicals products would be produced. In the previous section on lasing of molecules inside a filament, many lasing particles were the result of filament induced fragmentation/recombination of molecules; i.e. after chemical reactions. They are the formation of excited OH from water molecules [21], NH from breaking nitrogen and water molecules followed by recombination of N and H [22] and CH from the fragmentation of hydrocarbons [23], etc. Another example is the formation of nitric acid in humid air leading to the nucleation of H_2O–HNO_3 [33] inside the filament of a fs Ti-sapphire laser pulse. These molecules would act as seeds for filament induced condensation [34] and precipitation [35] in the atmosphere. See also a recent excellent review [36]. (More discussion will be given in Sect. 1.10.) A very recent example is the formation of hydrogenated amorphous carbon during filamentation of a Ti-sapphire laser in ethylene gas [37]. Such chemical reactions would give rise to stable yield because of intensity clamping.

There are other works on filament induced chemistry in liquid media. The first of such work was done in our laboratory [38]. H_2O_2 was generated by filamentation of the Ti-sapphire laser in water. Since then, there had been other work carried out in liquid media. For example, direct synthesis of nano-diamonds by femtosecond laser irradiation of ethanol was carried out recently by S. S. Yap et al. in Malaysia [39]. Nobuaki Nakashima and Tomoyuki Yatsuhashi of Osaka City University, Japan, have done much work on filament induced chemistry in liquids systematically. The following is adopted from a private communication with them.

Metal ions in solutions can be reduced by focused femtosecond laser pulses. The reduction of Eu^{3+} to Eu^{2+} in methanol was successfully observed for the first time in 2008 by this group [40]. The reaction was induced using focused 800 nm, 30 femtosecond Ti-sapphire laser pulses with a minimum laser energy of 100 nano-Joules/pulse, where the intensities were around the observation of supercontinuum generation [41]. This would mean filamentation because supercontinuum generation is a consequence of filamentation. In the strong field of the filament zone, solvent molecules are ionized and the electrons are captured by Eu^{3+} which reduces to Eu^{2+}. Other ions such as Sm^{3+}, Yb^{3+}, Fe^{3+} ions can be reduced to their corresponding doubly charged ions. For the cases of Ag^{+}, Au complex ion, and Pd^{2+}, their metal nanoparticles were produced [42]. These reactions would be useful for metal ion separation [41].

Moreover, carbon nanoparticles were synthesized by focusing femtosecond laser pulses into water in which organic molecules were mixed. In C_6H_6/water, fs Ti-sapphire laser filament enables C_6H_6 to react with hydroxyl radicals forming hydrophilic carbon nanoparticles without the aid of any additives (ex. surfactants) [43]. This is due to the formation of high density reactive species such as hydroxyl radicals which react with the benzene molecules in water [43]. Filament induced hydrated electrons [44, 45] in C_6F_6/water react with C_6F_6 forming fluorine-rich hydrophilic carbon nanoparticles [44]. The semi-ionic character of C–F bonds and the non-aggregating feature of the nanoparticles are the origins of water dispersibility acquired during the building-up process of nanoparticles in filaments. Similar experiment using CH_2Cl_2/water formed low-chlorine hydrophobic [45] carbon nanoparticles.

1.7 Remote Sensing and Vertical Propagation

Fluorescence is a delayed process depending on the lifetime of the fluorescing state if no collision is involved. This time is of the order of ns. If collision (a slower process than fluorescence) is involved, the whole process would take even a longer time. In filamentation, such fluorescence is easily detected. When a filament was formed in air in which other impurity gaseous molecules were mixed, not only ionization took place but also fluorescence was observed from the fragments of these impurity molecules. Some fluorescence had gain [46, 47]. When an air filament hit the surface of some solid chemicals or biological materials, fluorescence of some simple fragments (mostly diatomic) was also consistently observed [46, 47]. These so-called finger print fluorescence are the signatures of the gaseous and solid impurities/targets. Since a filament could in principle be projected to a long distance in air, one could carry out stand-off detection of almost all such impurities (pollutants or toxic materials). We have demonstrated such a sensing capability using only one laser, namely, the fs Ti-sapphire laser under filamentation and hitting targets in different phases, gas, solid and powder [46, 47]. The stand-off detection was from a distance of up to 100 m in both the laboratory environment and in the field of a cold winter (below −20 °C). These targets were the following: CH_4, C_2H_2 (gases), ethanol (vapor), powders of egg white, yeast, grain dusts of barley, corn and wheat;

water aerosols containing ($NaCl$, $PbCl_2$, $CuCl_2$ and $FeCl_2$), smoke from mosquito coil, metallic targets (lead, copper and aluminum). All of them showed distinctive finger print fluorescence.

In practice, there are some technical challenges. One is the well-known problem of multiple filamentation [2–7, 10]. In long distance detection, terawatt level laser pulses would have to be used. But the mode quality of such lasers is usually not good and not stable so that the filaments would fluctuate. Even if the mode quality were good and stable, one might have to deal with atmospheric turbulence that would disturb the filament pattern. As such, the return signal during a stand-off detection would be fluctuating. Quantitative measurement would be a big challenge.

The other challenge is the adaptation of the technique of propagation to the intrinsic physics of filamentation. If we send the laser pulse vertically into the atmosphere with the intention of using filamentation to detect pollutants or other materials at a certain altitude, one needs to be aware of the change of pressure vertically. This is because the clamped intensity is pressure independent [13]. Since the critical power for self-focusing P_c is given by [48]

$$P_c = \frac{3.77\lambda^2}{8\pi n_2 n_0} \tag{1.8}$$

where n_2 is the coefficient of Kerr nonlinear index of refraction, n_0, the linear index of refraction and λ, the wavelength. Since n_2 is proportional to the air density [49], from (1.8), P_c would have to increase at higher altitude where the pressure is lower. This would mean that at higher altitude, the input peak power of the laser pulse would have to increase. Since

$$I_{\text{clamped}} = \frac{P_{\text{clamped}}}{A} \propto \frac{P_c}{A} \tag{1.9}$$

where I_{clamped} and P_{clamped} are the clamped intensity and the corresponding clamped peak power of the laser pulse given a cross sectional area A of the filament. Since I_{clamped} does not change and P_c increases at higher altitude (lower pressure), (1.9) shows that the cross sectional area of the filament A has to increase thus increasing the filament volume. Essentially, the peak power or the energy content of the pulse would have to increase. Because the beam quality of a fs Ti-sapphire laser pulse at the multi-terawatt level is normally not perfect, increasing the power or energy content of the pulse at sea level would lead to multiple filamentation at sea level before propagating upward. Special care or design would have to be taken by the practitioner.

1.8 Industrial Application: Cutting Brittle Materials Using Filaments

This section deals the repetitive effect of using a train of high repetition fs or ps laser pulses to interact with a solid transparent and brittle material such as glass. A train of sufficiently powerful fs or ps pulses, when focused onto the surface of the material, will form filaments from the surface into the material as if they 'pierce' into the surface successively. These pulses are not powerful enough yet to form filaments in air. When a filament is formed inside the solid surface, a plasma will be created through multiphoton transition of electrons from the valence to the conduction band followed by inverse Bremsstrahlung and partial avalanche ionization [7]. After each interaction with a filament in the fs/ps time scale, the material will relax thermally in the ms time scale. But before the thermal relaxation of the material is complete; i.e. back to room temperature, a second filament will be formed in the material at the same position. This second filament will see the material modified by the first filament. The interaction physics will become different. In general, the preceding pulse will 'soften' the material for the second pulse to interact with. This thermal softening would mean less abruptness in the interaction, (hence, avoiding shockwave cracking of the material) thereby resulting in a clean 'cut'.

This technology is based upon the finding of a new class of interaction using high repetition rate (>100 kHz) pulse-train bursts of ps to fs laser pulses [50]. Each pulse-train burst contains tens of pulses with about 10 ps duration or shorter [51]. The pulses are spaced by a few ns. The focused pulses in each burst interact with the material from the surface inward. These pulses are spaced by a short time interval such that before the thermal and other slow relaxation inside the interaction zone have completed, the second pulse arrives. That is to say, each pulse prepares the material surface or bulk so as to alter the interaction with the subsequent pulse thereby improving or optimizing the interaction [51].

Based upon the above mentioned findings [50, 51], S. Abbas Hosseini added the advantage of fs laser filaments to efficiently cut brittle materials such as glass in a way that is superior to diamond cutting (see later). Because of intensity clamping inside the filament, the interaction is almost constant from shot to shot even if the laser energy fluctuates a bit. A variety of cutting techniques was thus invented by Hosseini [52–60] to deal with the change of materials and the change of cutting contours, shapes, etc. In no time, the technology spread into many industries that use brittle windows or surfaces. Almost all major companies such as Corning, Asahi, Siant Gobain and Schott, etc. that manufacture or use glass for automobile (car mirrors, dashboards), Aerospace (airplane windows), microelectronics, cell phones (cover windows) use this technology.

The technology is superior to diamond cutting from the economic and environmental point of view. The following is adapted from a private communication with S. Abbas Hosseini [61].

Glass cutting by diamond uses de-ionized (DI) water while doing the rolling. Micro cracks and chips could develop into major cracks and this reduces the yield.

Besides, to make multiple parts on a piece of glass, certain margins need to be kept so as to avoid cross cracking. For example, on a piece of glass whose size can accommodate 7 mirrors, only 5 can be cut by diamond because of the margins between two mirrors. Finally glass needs to get grinded by 300–400 μm to remove micro cracks from the edges. This also uses a lot of machinery and DI water. As to the quality, let us look at an example. Soda lime glass with a thickness of 1.6 mm will have 30–40 MPa bending strength right after diamond cut. This test is done by a machine called 4-bend testing. It uses two base rods as glass holder then two other rods from the top to push the glass to bend. The pressure at which the glass cracks is measured. This number is not acceptable to Motor vehicle authorities. They thus need to grind the edges to remove the cracks. After this process, they reach 70–80 MPa. Using fs laser filament cutting, without using water, a cutting speed at 500 mm/s is used; it is twice faster than diamond rollers. The edges are free from micro cracks and chips; therefore, it doesn't need grinding and that means saving water. Bending test results in 120–130 MPa. Now if only 4–5 μm of the sharp edge gets dry grinded (no water and just one grinder touching the glass edges to chamfer it) it will result in 220 MPa! Also, on the same piece of glass whose size can accommodate 7 mirrors, 7 parts can be cut instead of 5. The yield jumps from 70 to 99.9%; that is a huge glass saving. This helps the reduction of CO_2 emission in glass production.

In another aspect, cutting glass less than 500 μm thick with diamond was a big challenge up until 2012 because thinner glass couldn't resist the diamond roller pressure. No one could cut 400 μm thick glass. Previously, using diamond cutting, 4 layers of glass each 550 μm thick are used on a cell phone. Using filament cutting technique, each layer becomes 150–200 μm thick. Suddenly the total thickness drops by about 800 μm or becomes about three times thinner. That means about 63% reduction in glass production; this also means less CO_2 emission. Also, the yield in cutting is better; i.e. more saving in the cost. This is why smart phones get thinner, lighter and cheaper.

1.9 Wood Cutting

Similar to cutting brittle materials discussed in Sect. 1.8, an air filament was used to cut wood surface there-by exposing the cell structures under an electron microscope [62, 63]. The intense light in the filament zone exploded the cellulose material 'instantaneously' without oxidation (i.e. no burning). Each cell's interior structure was exposed. This 'clean' cut would never be achieved using even the sharpest mechanical knife to cut the wood. Mechanical cutting will always result in a significant squeezing of the wood cells thus blocking the view of the interior of the cell [63].

The physics of interaction has not been studied systematically by the author's group and probably not by other groups to the limited knowledge of the author. A speculation is the following. Similar to the case of glass or metal, multiphoton transition from the valence to conduction band followed by inverse Bremsstrahlung and a partial cascade/avalanche ionization [7] would have taken place resulting in a local plasma within the laser's pulse duration; i.e. the plasma was created almost instantaneously. The plasma would expand quickly and explode the wood material. The explosion has generated a lot of wood dust particles flying out of the surface [64]. This would mean that the flying wood particles would carry away most of the energy in the plasma in the form of kinetic energy. Hence, not much heat would be left on the surface; i.e. 'cold' cutting. The 'cold' surface would not interact with ambient oxygen; hence, no oxidation or charring of the wood was observed [62–64].

1.10 Filament Induced Discharge Guiding, Condensation, Precipitation and Air Motion (Wind)

During the 'instantaneous' excitation events (fs time scale) in a filament in air (see Sect. 1.2), energy is fed from the laser pulse into the interaction processes and get 'stored' temporarily in the filament zone. Soon after this, a series of relaxations take place in the sense that the excitation energy will be released. Radiative relaxation of molecules will emit radiations out of the system. Collisions in events such as plasma recombination, rotational relaxation, chemical reaction, and so on take place at different times. The fast relaxation in the fs, ps and ns time scale has led to the phenomena and applications such as fluorescence, harmonics generation, 4WM, molecular alignment, etc. Some phenomena such as molecular alignment through the excitation of rotational wave packets [9] and the generation of THz wave due to plasma oscillation and 4WM [65] were not discussed.

Meanwhile, the recombination of the plasma and the relaxation of all the excited molecules would result in a sudden non-uniform heating resulting in a very high temperature (more than one thousand degrees Kelvin) in the filament's axial zone while near the filament's surface, it was around ambient temperature [66]. This high temperature gradient would give rise to a sudden expansion of the filament zone in the microsecond time scale [66]. This would result in a shock wave propagating outward from the filament. The shock wave would propagate and decay into an acoustic wave in the microsecond to millisecond time scale. The filament zone would become a thermally hot zone. The pressure in this hot zone would be lower than the ambient pressure [66]. This low pressure zone had been shown to be physically responsible for guiding discharge in air in relation to the dream of using filament to guide lightning [67]. This is because the electrical conductivity of the low pressure zone is high. It would thus become a discharge channel in a high voltage zone guiding corona discharge [68–70]. An excellent recent review is given by J.-P. Wolf on using short laser pulse for weather control [36].

The pressure in the low pressure hot zone would return to ambient pressure in a few microseconds [71]. Its size would expand to a few times the original size in about 500 μs [66]. This expanded hot zone would relax thermodynamically; i.e. convection would follow. Convection means heat conduction together with gas motion and mixing. The time scale could last up to 90 ms [66]. Air flow would occur; i.e. wind would be induced by filamentation.

The thermodynamic relaxation of the filament zone described in the preceding paragraph would have practical consequences depending on the environment in which the filament is formed and depending on the performance of the laser. For example, if a high repetition rate (1 kHz, for example) fs Ti-sapphire laser were used to generate filaments inside a cloud chamber, the resultant air motion would become a turbulence [35, 72, 73]. In these experiments [35, 72–75], the humidity inside the cloud chamber (0.5 m × 0.5 m × 0.2 m) was high (saturated or sub-saturated). The bottom plate was at a temperature of −46 °C while at the top, room temperature was maintained [35]. The large temperature gradient plus the air turbulence would induce condensation and precipitation [34–36, 72–75]. This is because the turbulence would mix up the moist air with a large temperature gradient such that after mixing, super-saturation would result in the region around the filament. Chemicals such as HNO_3–H_2O as well as the background impurities would act as cloud condensation nuclei and grow in the super-saturated environment. At high repetition rate, a quasi-continuous build-up of the condensation would result in the formation of large size water/ice droplets that would fall down to the bottom of the cloud chamber [74]. If the repetition rate were slow (e.g. 10 Hz), only condensation would be seen because the growth of the water droplets into large size particles could not be sustained at low repetition rate of excitation. Condensation had been observed not only in a cloud chamber but also in the field (see [36] and references therein). Filamentation induced precipitation could be used as an alternative way to remove environmental pollutants.

1.11 Conclusion

This paper gives an overview of a limited number of applications/phenomena arising at different stages of the temporal evolution of a fs Ti-sapphire laser filament (mostly in air). The mechanism of these applications/phenomena vary from quantum physics in the short time scale (fs) through chemistry in the intermediate time scale (ps to ns) to classical physics in the long time scale (μs to s). It is hoped that the brief analysis would stimulate new ideas for the discovery of new phenomena and applications.

Acknowledgements The author would like to thank Prof. Nobuaki Nakashima and Prof. Tomoyuki Yatsuhashi of Osaka City University, Japan, Dr. S. Abbas Hosseini of Citrogene, Calif., USA and Prof. Olga Kosareva of Moscow State University for generously sharing their scientific knowledge with the author. This work was supported by the Center for Optics, Photonics and Laser (COPL), Laval University, Quebec City, Canada.

References

1. A. Braun, G. Korn, X. Liu, D. Du, J. Squier, G. Mourou, Opt. Lett. **20**, 73 (1995)
2. S.L. Chin, S.A. Hosseini, W. Liu, Q. Luo, F. Theberge, N. Aközbek, A. Becker, V.P. Kandidov, O.G. Kosareva, H. Schroeder, Can. J. Phys. **83**, 863 (2005)
3. A. Couairon, A. Mysyrowicz, Phys. Rep. **44**, 47 (2007)
4. L. Bergé, S. Skupin, R. Nuter, J. Kasparian, J.-P. Wolf, Rep. Prog. Phys. **70**, 1633 (2007)
5. J. Kasparian, J.-P. Wolf, Opt. Express **16**, 466 (2008)
6. V.P. Kandidov, S.A. Shlenov, O.G. Kosareva, Quantum Electron. **39**, 205 (2009)
7. S.L. Chin, *Femtosecond Laser Filamentation* (Springer, New York, 2010)
8. A. Talebpour, J. Yang, S.L. Chin, Semi-empirical model for the rate of tunnel ionization of N_2 and O_2 molecule in an intense Ti: Sapphire laser pulse. Opt. Commun. **162**, 29–32 (1999)
9. E.A. Volkova, A.M. Popov, O.V. Tikhonova, Polarisation response of a gas medium in the field of a high-intensity ultrashort laser pulse: high order Kerr nonlinearities or plasma electron component? Quantum Electron. **42**(8), 680–686 (2012)
10. S.L. Chin, T.-J. Wang, C. Marceau, J. Wu, J.S. Liu, O. Kosareva, N. Panov, Y.P. Chen, J.-F. Daigle, S. Yuan, A. Azarm, W.W. Liu, T. Seideman, H.P. Zeng, M. Richardson, R. Li, Z.Z. Xu, Advances in intense femtosecond laser filamentation in air. Laser Phys. **22**, 1 (2012)
11. J. Kasparian, R. Sauerbrey, S.L. Chin, The critical laser intensity of self-guided light filaments in air. Appl. Phys. B **71**, 877–879 (2000)
12. F. Théberge, W. Liu, P.T. Simard, A. Becker, S.L. Chin, Plasma density inside a femtosecond laser filament in air: strong dependence on external focusing. Phys. Rev. E **74**, 036406 (2006)
13. **a**. J. Bernhardt, W. Liu, S.L. Chin, R. Sauerbrey, Pressure independence of intensity clamping during filamentation: theory and experiment. Appl. Phys. B **91**, 45 (2008); **b**. S. Hosseini, O. Kosareva, N. Panov, V.P. Kandidov, A. Azarm, J.-F. Daigle, A.B. Savel'ev, T.-J. Wang, S.L. Chin, Femtosecond laser filament in different air pressures simulating vertical propagation up to 10 km. Laser Phys. Lett. **9**, 868 (2012)
14. W. Liu, S.L. Chin, Abnormal wavelength dependence of the self-cleaning phenomenon during femtosecond-laser-pulse filamentation. Phys. Rev. A **76**, 013826 (2007)
15. Q. Luo, W. Liu, S.L. Chin, Lasing action in air induced by ultrafast laser filamentation. Appl. Phys. B **76**, 337 (2003)
16. **a**. H.L. Xu, A. Azarm, J. Bernhardt, Y. Kamali, S.L. Chin, The mechanism of nitrogen fluorescence inside a femtosecond laser filament in air. Chem. Phys. **360**, 171 (2009); **b**. S. Owada, A. Azarm, S. Hosseini, A. Iwasaki, S.L. Chin, K. Yamanouchi, Amplified spontaneous $C^3\Pi_u$-$B^3\Pi_g$ emission and rotational and vibrational state distributions in $C^3\Pi_u$ state of N_2 in femtosecond laser induced filament in air. Chem. Phys. Lett. **581**, 21 (2013)
17. Helong Li, Mengyao Hou, Hongwei Zang, Fu Yao, Erik Lötstedt, Toshiaki Ando, Atsushi Iwasaki, Kaoru Yamanouchi, Xu Huailiang, Significant enhancement of N_2^+ lasing by polarization-modulated ultrashort laser pulses. Phys. Rev. Lett. **122**, 013202 (2019)
18. A. Becker, A.D. Bandrauk, S.L. Chin, S-matrix analysis of non-resonant multiphoton ionisation of inner-valence electrons of the nitrogen molecule. Chem. Phys. Lett. **343**, 345–350 (2001)
19. A. Dogariu, J.B. Michael, M.O. Scully, R.B. Miles, Science **331**, 442 (2011)
20. A. Laurain, M. Scheller, P. Polynkin, Phys. Rev. Lett. **113**, 253901 (2014)
21. S. Yuan, T.J. Wang, Y. Teranishi, A. Sridharan, S.H. Lin, H.P. Zeng, S.L. Chin, Lasing action in water vapor induced by ultrashort laser filamentation. Appl. Phys. Lett. **102**, 224102 (2013)
22. S. Yuan, T.J. Wang, P. Lu, S.L. Chin, H.P. Zeng, Humidity measurement in air using filament-induced nitrogen monohydride fluorescence spectroscopy. Appl. Phys. Lett. **104**, 091113 (2014)
23. S. Hosseini, A. Azarm, J.-F. Daigle, Y. Kamali, S.L. Chin, Filament-induced amplified spontaneous emission in air-hydrocarbons gas mixture. Opt. Commun. **316**, 61 (2014)
24. Y. Hatano, Phys. Rep. **313**, 109 (1999)
25. F. Kong, Q. Luo, H. Xu, M. Sharifi, D. Song, S.L. Chin, Explosive photodissociation of methane induced by ultrafast intense laser. J. Chem. Phys. **125**, 133320 (2006)

26. F. Kong, S.L. Chin, Non-coulomb explosions of molecules in intense laser fields, in *Progress in Ultrafast Intense Laser Science III*, Springer series in chemical physics, ed. by K. Yamanouchi, S.L. Chin, P. Agostini, G. Ferrante, vol. 3, Chap. 6, p. 113 (2008)
27. A. Azarm, H.L. Xu, Y. Kamali, J. Bernhardt, D. Song, A. Xia, Y. Teranishi, S.H. Lin, F. Kong, S.L. Chin, Direct observation of super-excited states in methane created by a femtosecond intense laser field. J. Phys. B **41**, 225601 (2008)
28. S.L. Chin, H.L. Xu, Population inversion in fluorescing fragments of super-excited molecules inside an air filament. Chin. Phys. B **24**, 013301 (2015)
29. S.L. Chin, A. Azarm, H.L. Xu, T.J. Wang, M. Sharifi, A. Talebpour, Experiments in population trapping in atoms and molecules by an intense short laser pulse, in *Progress in Ultrafast Intense Laser Science VIII*, ed. by K. Yamanouchi, M. Nisoli, W.T. Hill III (Springer, Heidelberg, 2012), pp. 79–96
30. A. Azarm, S.M. Sharifi, A. Sridharan, S. Hosseini, Q.Q. Wang, A.M., O.V. Tikhonova, E.A. Volkova, S.L. Chin, Population trapping in Xe atoms. J. Phys.: Conf. Ser. **414**, 012015 (2013)
31. **a**. N. Aközbek, A. Iwasaki, A. Becker, M. Scalora, S.L. Chin, C.M. Bowden, Third-harmonic generation and self-channeling in air using high-power femtosecond laser pulses. Phys. Rev. Lett. **89**, 143901 (2002); **b**. F. Théberge, J. Filion, N. Aközbek, Y. Chen, A. Becker, S.L. Chin, Self-stabilization of third-harmonic pulse during two-color filamentation in gases. Appl. Phys. B **87**, 207 (2007)
32. F. Théberge, N. Aközbek, W. Liu, A. Becker, S.L. Chin, Tunable ultrashort laser pulses generated through filamentation in gases. Phys. Rev. Lett. **97**, 023904 (2006)
33. Y. Petit, S. Henin, J. Kasparian, J.P. Wolf, Appl. Phys. Lett. **97**, 021108 (2010)
34. P. Rohwetter, J. Kasparian, K. Stelmaszczyk, Z. Hao, S. Henin, N. Lascoux, W.M. Nakaema, Y. Petit, M. Queißer, R. Salamé, E. Salmon, L. Wöste, J.-P. Wolf, Nat. Photonics **4**, 451 (2010)
35. J. Ju, J.S. Liu, C. Wang, H. Sun, W. Wang, X. Ge, C. Li, S. L. Chin, R. Li, Z. Xu, Laser-filamentation-induced condensation and snow formation in a cloud chamber. Opt. Lett. **37**, 1214 (2012)
36. J.P. Wolf, Short-pulse lasers for weather control. Rep. Prog. Phys. **81**(2), 026001 (2018)
37. A. Matsuda, T. Hayashi, R. Kitaura, A. Hishikawa, Femtosecond laser filamentation in gaseous ethylene: formation of hydrogenated amorphous carbon. Chem. Lett. **46**, 1426–1429 (2017). https://doi.org/10.1246/cl.170613
38. S.L. Chin, S. Lagacé, Generation of H_2, O_2 and H_2O_2 from water using intense femtosecond laser pulses and the possibility of application to sterilization. Appl. Opt. **35**, 907 (1996)
39. C.-H. Nee, S.-L. Yap, T.-Y. Tou, H.-C. Chang, S.-S. Yap, Direct synthesis of nanodiamonds by femtosecond laser irradiation of ethanol. Sci. Rep. **6**, 33966 (2016). https://doi.org/10.1038/srep33966
40. D. Nishida, M. Kusaba, T. Yatsuhashi, N. Nakashima, Reduction of Eu^{3+} to Eu^{2+} by an intense femtosecond laser pulse in solution. Chem. Phys. Lett. **465**, 238–240 (2008)
41. N. Nakashima, T. Yatsuhashi, K. Sakota, I. Iwakura, K. Yokoyama, Laser chemistry for Eu recovery, in *The 98th CSJ Annual Meeting* (2018), 1PB-039
42. **a**. N. Nakashima, K. Yamanak, M. Saeki, H. Ohba, S. Taniguchi, T. Yatsuhashi, Metal ion reductions by femtosecond laser pulses with micro-Joule energy and their efficiencies. J. Photochem. Photobiol. A: Chem. **319**, 70–77 (2016); **b**. N. Nakashima, K. Yamanaka, T. Yatsuhashi, Ionic valence change of metal ions in solution by femtosecond laser excitation accompanied by white-light laser. Chin. J. Phys. **52**, 504–518 (2014)
43. **a**. T. Yatsuhashi, N. Uchida, K. Nishikawa, Novel method of producing carbon nanoparticles on benzene/water interface with femtosecond laser plasma filament. Chem. Lett. 2012, **41**(7), 722–724; **b**. T. Hamaguchi, T. Okamoto, K. Mitamura, K. Matsukawa, T. Yatsuhashi, Synthesis of hydrophilic and hydrophobic carbon nanoparticles from benzene/water bilayer solution with femtosecond laser generated plasma filaments in water. Bull. Chem. Soc. Jpn. 2015, **88**(2), 251–261
44. T. Okamoto, K. Mitamura, T. Hamaguchi, K. Matsukawa, T. Yatsuhashi, Synthesis of fluorine-doped hydrophilic carbon nanoparticles from hexafluorobenzene by femtosecond laser pulses. Chem. Phys. Chem. **18**(9), 1007–1011 (2017)

45. T. Okamoto, E. Miyasaka, K. Mitamura, K. Matsukawa, T. Yatsuhashi, Precipitation of dichloromethane as low-chlorine carbon nanoparticles from water by femtosecond laser pulses. J. Photochem. Photobiol., A **344**, 178–183 (2017)
46. S.L. Chin, H.L. Xu, Q. Luo, F. Théberge, W. Liu, J.F. Daigle, Y. Kamali, P.T. Simard, J. Bernhardt, S.A. Hosseini, M. Sharifi, G. Méjean, A. Azarm, C. Marceau, O. Kosareva, V.P. Kandidov, N. Aközbek, A. Becker, G. Roy, P. Mathieu, J.R. Simard, M. Châteauneuf, J. Dubois, Filamentation "remote" sensing of chemical and biological agents/pollutants using only one femtosecond laser source. Appl. Phys. B **95**, 1 (2009)
47. H.-L. Xu, S.L. Chin, Femtosecond laser filamentation for atmospheric sensing. Sensors **11**, 32 (2011)
48. J.H. Marburger, Prog. Quant. Electron. **4**, 35 (1975)
49. P. Sprangle, J.R. Peñano, B. Hafizi, Phys. Rev. E **66**, 046–418 (2002)
50. M. Lapczyna, K.P. Chen, P.R. Herman, H.W. Tan, R.S. Marjoribanks, Ultra high repetition rate (133 MHz) laser ablation of aluminum with 1.2-ps pulses. Appl. Phys. A **69**(1), S883–S886 (1999)
51. P.R. Herman, R. Majoribanks, A. Oettl, Burst-ultrafast laser machining method. US patent no. US 6,552,301 B2, April 22, 2003
52. S. Abbas Hosseini, Method of closed form release for brittle materials using burst ultrafast laser pulses, US patent application publication, Pub. No.: US 2015/0136743 A1, May 21, 2015
53. S. Abbas Hosseini, Method and apparatus for laser processing of silicon by filamentation of burst ultrafast laser pulses, US Patent application publication, Pub. No.: US 2015/0151380 A1, Jun. 4, 2015
54. S. Abbas Hosseini, R. Myerhofer, Method and system for scribing heat processed transparent materials, US patent application publication, Pub. No.: US 2016/0318790 A1, Nov. 3, 2016
55. S. Abbas Hosseini, Method and apparatus for spiral cutting a glass tube using filamentation by burst ultrafast laser pulses, US patent application publication, Pub.: US 2015/0140241 A1, May 21, 2015
56. S. Abbas Hosseini, Method and apparatus for machining diamonds and gemstones using filamentation by burst ultrafast laser pulses, US patent application publication, Pub.: US 2015/0121960 A1, May 7, 2015
57. S. Abbas Hosseini, Method of fabricating a glass magnetic hard drive disk platter using filamentation by burst ultrafast laser pulses, US patent application publication, Pub.: US 2015/0118522 A1, April 30, 2015
58. S. Abbas Hosseini, Method and apparatur for non-ablative, photoacoustic compression machining in transparent materials using filamentation by burst ultrafast laser pulses, US patent, Patent No.: US 9,102,011 B2, Aug. 11, 2015
59. S. Abbas Hosseini, Method and apparatur for performing laser filamentation within transparent materials, US patent, Patent No.: US 9,102,007 B2, Aug. 11, 2015
60. S. Abbas Hosseini, Method of material processing by laser filamentation, US patent application publication, Pub.: US 2013/0126573 A1, May 23, 2013
61. S. Abbas Hosseini, Citrogene. www.citrogene.com
62. S.L. Chin, N. Naderi, S. Lagace, Use of ultrafast intense laser for processing lignocellulosic material. United States Patent, 5862845 (1999)
63. N. Naderi, S. Lagacé, S.L. Chin, Preliminary investigations of ultrafast intense laser wood processing. Forest Prod. J. **49**, 72–76 (1999)
64. F. Théberge, S. Petit, A. Iwasaki, M.R. Kasaai, S.L. Chin, Ultrafast intense laser "explosion" of hardwood. Appl. Surf. Sci. **191**, 328–333 (2002)
65. V.A. Andreeva, O.G. Kosareva, N.A. Panov, D.E. Shipilo, P.M. Solyankin, M.N. Esaulkov, P. Gonzalez de Alaiza Martinez, A.P. Shkurinov, V.A. Makarov, L. Bergé, S.L. Chin, Ultrabroad terahertz spectrum generation from an air-based filament plasma. Phys. Rev. Lett. **116**, 063902 (2016)
66. G. Point et al., Generation of long-lived underdense channels using femtosecond filamentation in air. J. Phys. B **48**, 094009 (2015)

67. S. Tzortzakis, B. Prade, M. Franco, A. Mysyrowicz, Femtosecond laser-guided electric discharge in air. Phys. Rev. E **64**, 057401 (2001)
68. T.-J. Wang, Y. Wei, Y. Liu, N. Chen, Y. Liu, J. Ju, H. Sun, C. Wang, H. Lu, J. Liu, S.L. Chin, R.X. Li, Z.Z. Xu, Direct observation of laser guided corona discharges. Sci. Rep. **5**, 18681 (2015)
69. Y. Liu, T. Wang, N. Chen, S. Du, J. Ju, H. Sun, C. Wang, J. Liu, H. Lu, S.L. Chin, R. Li, Z. Xu, Z. Wang, Probing the effective length of plasma inside a filament. Opt. Express **25**, 11078 (2017)
70. Y.X. Wei, Y.X. Liu, T.-J. Wang, S.L. Chin, R.X. Li, Spectroscopic analysis of high electric field enhanced ionization in laser filaments in air for corona guiding. High Power Laser Sci. Eng. **4**, e8 (2016)
71. G. Point, Private communication (2015)
72. J.J. Ju, H.Y. Sun, A. Sridharan, T.-J. Wang, C. Wang, J.S. Liu, R.X. Li, Z.Z. Xu, S.L. Chin, Laser-filament-induced snow formation in a subsaturated zone in a cloud chamber: Experimental and theoretical study. Phys. Rev. E **88**, 062803 (2013)
73. H.Y. Sun, J.S. Liu, C. Wang, J.J. Ju, Z.X. Wang, W.T. Wang, X.C. Ge, C. Li, S.L. Chin, R.X. Li, Z.Z. Xu, Laser filamentation induced air-flow motion in a diffusion cloud chamber. Opt. Express **21**, 9255 (2013)
74. J.J. Ju, J.S. Liu, H. Liang, Y. Chen, H.Y. Sun, Y.H. Liu, J.W. Wang, C. Wang, T.J. Wang, R.X. Li, Z.Z. Xu, S.L. Chin, Femtosecond laser filament induced condensation and precipitation in a cloud chamber. Sci. Rep. **6**, 25417 (2016)
75. J.J. Ju, T. Leisner, H.Y. Sun, A. Sridharan, T.J. Wang, J.W. Wang, C. Wang, J.S. Liu, R.X. Liu, Z.Z. Xu, S.L. Chin, Laser-induced supersaturation and snow formation in a sub-saturated cloud chamber. Appl. Phys. B **117**, 1001 (2014)

Chapter 2
Mechanism of Population Inversion in N_2^+

Youyuan Zhang, Erik Lötstedt, and Kaoru Yamanouchi

Abstract When intense femtosecond laser pulses are focused in air, unidirectional and coherent radiation called air-lasing is generated, originating from the population inversion processes between an electronically excited state and the electronic ground state of N_2 and N_2^+. In 2015, it was shown experimentally that the air-lasing at 391 nm, corresponding to the $B^2\Sigma_u^+(v = 0) - X^2\Sigma_g^+(v = 0)$ emission of N_2^+, is generated in an sub-10 fs near-IR laser field (Xu et al, Nat Commun 6:8347, 2015 [6]). The behavior that N_2^+ generated in the pulse is exposed to the intense laser field immediately after its birth, is considered critical in leading to the population inversion between the B and X states. This chapter includes the demonstrations that the sudden exposure to an intense laser field could result in efficient population transition in a two-level system at off-resonance (Zhang et al, J Phys B: At Mol Opt Phys 50:185603, 2017 [32]), and that population inversion can be achieved in $N_2^+$by using sudden turn-on pulse (Zhang et al, J Phys B: At Mol Opt Phys 52:055401, 2019 [33]). This scenario of the population transfer to the excited state is expected to be universal and can be applied to an interpretation of population inversion of any kind of atomic and molecular ions created in a pulsed intense laser field.

2.1 Introduction

The concept of the standoff spectroscopy was introduced first in [1]. In 2011, a simple approach of standoff spectroscopy, where the backward-directed lasing from air, mostly from ionized oxygen molecules and ionized nitrogen molecules, was proposed [2]. Since then, the air-lasing has been regarded as a promising phenomenon, with

---

Y. Zhang · E. Lötstedt · K. Yamanouchi (✉)
Department of Chemistry, School of Science, The University of Tokyo,
7-3-1 Hongo, Bunkyo-ku, Tokyo 113-0033, Japan
e-mail: kaoru@chem.s.u-tokyo.ac.jp

Y. Zhang
e-mail: youyuan.zhang@chem.s.u-tokyo.ac.jp

E. Lötstedt
e-mail: lotstedt@chem.s.u-tokyo.ac.jp

K. Yamanouchi and D. Charalambidis (eds.), *Progress in Ultrafast Intense Laser Science XV*, Topics in Applied Physics 136,
https://doi.org/10.1007/978-3-030-47098-2_2

which pollutants in the atmosphere can be effectively detected. In the past few years, it has been reported that by propagating an IR pump pulse in air, UV emission will be generated from filamentation in air [3, 4]. Among the coherent emissions generated from the laser induced filamentation in air, the forward and backward emission at 391 nm achieved by near-IR femtosecond laser pulses is reported to be the emission between the vibrational ground states of the electronically excited $B^2\Sigma_u^+$ state and electronic ground $X^2\Sigma_g^+$ state of N_2^+ [5–9], which requires the higher probability of N_2^+ to be prepared in $B^2\Sigma_u^+$ than in $X^2\Sigma_g^+$, leading to the population inversion. For further application, especially in enhancing the emission signal, we need to understand the mechanism of the population inversion in N_2^+.

2.1.1 Superradiant Emission Mechanism

Through the measurement and analysis of the temporal profiles of the coherent emission, the signatures of the superradiance was reported in [10]. Since then, the superradiance, where a group of excited molecules interact with light collectively and coherently, has been suggested as a possible mechanism of air-lasing [10–13]. Meanwhile, it has been demonstrated that seed-triggered superradiance can be enhanced by population inversion achieved in N_2^+ [12], and a mechanism of stimulated-emission-induced seed amplification was suggested in [6, 14, 15].

2.1.2 Excitation Mechanism

While the emission process is still controversial, the excitation process from electronically ground $X^2\Sigma_g^+$ state to the electronically excited $B^2\Sigma_u^+$ state of N_2^+ becomes clearer. The field-induced inelastic recollision process was suggested as the excitation mechanism [11, 16], but later, the recollision process was reported to be not related to the air-lasing phenomenon [17, 18]. In [6, 8, 9, 15, 19–21], the strong-field ionization induced efficient optical coupling among the three electronic states, $X^2\Sigma_g^+$, $A^2\Pi_u$, and $B^2\Sigma_u^+$ states of N_2^+ shown in Fig. 2.1 as well as the important role of the electronic $A^2\Pi_u$ state in depleting population in $X^2\Sigma_g^+$ were proposed.

In [22, 23], it was demonstrated that the alignment angle between the molecular axis of N_2^+ and the polarization direction of the laser pulse affects the population transfer. Besides the vibrational population inversion, the rotational coherence in the 391 nm air-lasing was observed by pump-probe measurements [24], the coherent coupling between rotational states of $X^2\Sigma_g^+$ and those of $B^2\Sigma_u^+$ in the strong laser field was shown by time-resolved spectroscopy [25], and interpreted theoretically [26], and the rotational excitation was suggested as a possible mechanism of population inversion [27–29]. It was also reported based on spectroscopic measurements that population inversion can be achieved between specific rotational states in the

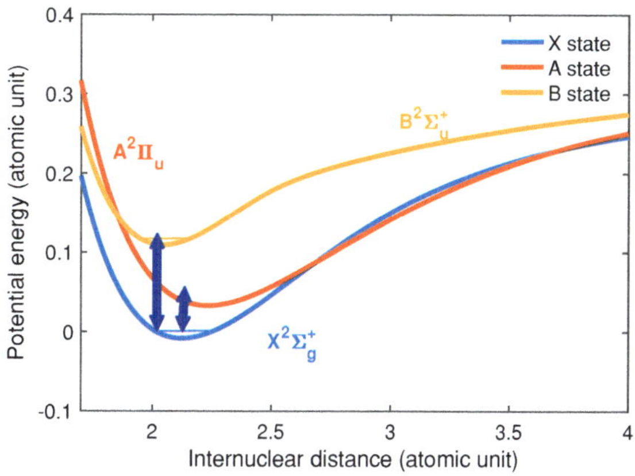

Fig. 2.1 Potential energy curves of the $X^2\Sigma_g{}^+$, $A^2\Pi_u$, and $B^2\Sigma_u{}^+$ states of $N_2{}^+$. The transitions between electronic states are shown by deep-blue double arrows

$X^2\Sigma_g{}^+$ state and the $B^2\Sigma_u{}^+$ state by the irradiation of a 800 nm laser pulse [27] as well as by the irradiation of a 1500 nm laser pulse [28], leading to air-lasing without net population inversion between the vibrational states.

2.1.3 *Efficient Excitation in a Two-Level System at Off-Resonance*

In order to understand the complex population inversion process in $N_2{}^+$, we need to establish theoretical model. According to the time-dependent generation rate of $N_2{}^+$ in an intense laser field, the ionization process preferentially occurs in the peak region of each optical cycle. As long as the ionization time is short enough, the electronic excitation of $N_2{}^+$ can be simulated from the time immediately after the ionization, represented by a sudden turn-on laser pulse interacting with $N_2{}^+$. The electronic ground X state of $N_2{}^+$ is coupled optically with the electronically excited B state by a light field component parallel to the N–N molecular axis while the X state is coupled optically with the electronically excited A state by a light field component perpendicular to the molecular axis. On the other hand, there is no optical coupling between the A state and the B state because both have *ungerade* symmetry [30, 31]. Therefore, the two coupling processes can be considered separately as a two-level system in theoretical investigation. When the $N_2{}^+$ is aligned parallel to the polarization direction of the laser pulse, only $X^2\Sigma_g{}^+$ and $B^2\Sigma_u{}^+$ need to be considered, resulting in an optically coupled two-level system. In [32], we analyzed theoretically a response of a two-level system to a sudden turn-on laser pulse, and demonstrated the efficient population transfer in an off-resonant case.

The dynamics of a laser-driven two-level system is given by the time-dependent Schrödinger equation (2.1):

$$i\hbar\frac{\partial}{\partial t}\Psi = H\Psi, \tag{2.1}$$

where the time-dependent Hamiltonian H is

$$H(t) = \begin{pmatrix} \varepsilon_1 & -\mu E(t) \\ -\mu E(t) & \varepsilon_2 \end{pmatrix}, \tag{2.2}$$

where ε_i is the eigenenergy of the i-th level, μ is the dipole moment between the two levels, and $E(t)$ is the laser field defined as

$$E(t) = \mathcal{E}_0 g(\tau) \cos(\tau), \tag{2.3}$$

where $\mathcal{E}_0$ is the peak field strength and $g(\tau)$ is the envelope function of the sudden turn-on pulse,

$$g(\tau) = \begin{cases} 0 & \text{if } \tau < 0, \\ e^{-\frac{\tau^2}{\sigma^2}} & \text{if } \tau \geq 0, \end{cases} \tag{2.4}$$

where σ determines the pulse width and $\tau = \omega t$ with ω being the laser frequency.

In order to examine the sudden turn-on behavior, we modify the pulse shape using an envelope function $g_k(\tau)$ given with a steepness parameter k as

$$g_k(\tau) = \begin{cases} e^{-\frac{\tau^2}{\sigma^2}-\frac{\tau^2}{k^2}} & \text{if } \tau < 0, \\ e^{-\frac{\tau^2}{\sigma^2}} & \text{if } \tau \geq 0. \end{cases} \tag{2.5}$$

When $k \ll 1$, the pulse becomes close to a sudden turn-on pulse, and when $k \gg 1$, the pulse becomes a full Gaussian pulse.

The final population in the excited state at the two-photon resonance after the laser pulse vanishes is plotted as a function of the coupling strength parameter $a = -\frac{\mu \mathcal{E}_0}{2\hbar\omega}$, where $\hbar\omega$ represents the one photon energy in laser pulse.

As shown in Fig. 2.2, when the pulse loses the sudden turn-on behavior at $k = 10$, population inversion requires a larger value of $a \approx 0.89$, which is larger than $a = 0.6$ required in the case of a sudden turn-on pulse at $k = 0.01$.

2.1.4 *Population Inversion Achieved by Sudden Turn-On Pulse in N_2^+*

In [6, 15], using a sudden turn-on intense laser pulse, a numerical simulation of population transfer in N_2^+ generated by intense laser pulse was performed and the population inversion was shown to be achieved between the $X^2\Sigma_g^+$ state and the $B^2\Sigma_u^+$ state of N_2^+ with the same laser parameters as in their experiment. Later in [33], by quasi-stationary Floquet theory with a sudden turn-on pulse, we investigated the mechanism of the population inversion achieved between the $X^2\Sigma_g^+$ state and the $B^2\Sigma_u^+$ state of N_2^+ through electronic and vibrational excitations induced by

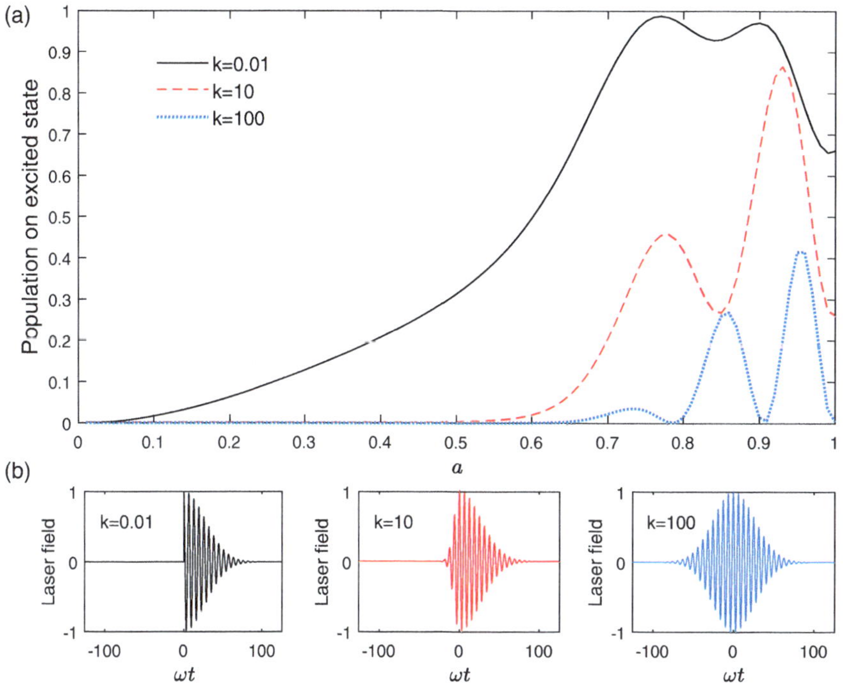

Fig. 2.2 **a** The final population in the excited state plotted as a function of the coupling strength parameter a. The initial conditions are 100% population in the ground level at $t = -\infty$. **b** The laser electric fields $g_k(\tau)\cos(\tau)$ for $k = 0.01$, $k = 10$, and $k = 100$

the laser pulse and revealed the important role of the $A^2\Pi_u$ state in reducing the inversion threshold between the $X^2\Sigma_g{}^+$ state and the $B^2\Sigma_u{}^+$ state depending on the alignment angle of the N-N axis with respect to the laser polarization direction.

While the vibrational inversion has been widely investigated, it was also suggested that the rotational degrees of freedom can play a certain role in achieving the lasing at 391 nm on the basis of the rotational coherence identified by the pump-probe measurements [24]. The coherent couplings between the rotational levels of the $B^2\Sigma_u{}^+$ state and those of the $X^2\Sigma_g{}^+$ state were identified by time-resolved spectroscopy [25] and were theoretically interpreted [26]. It has also been suggested that the rotational excitation is one of the possible mechanisms in achieving population inversion between the $B^2\Sigma_u{}^+$ state and the $X^2\Sigma_g{}^+$ state [27, 28]. It was shown that the population inversion can be achieved between rotational levels in the $B^2\Sigma_u{}^+$ state and those in the$X^2\Sigma_g{}^+$ state when the wavelengths of the excitation laser were 800 nm [27, 34] and 1500 nm [28] by spectroscopic measurements.

The simulation of rotational excitation in a rigid rotor model for molecular alignment and orientation has been established [35–38]. However, as for $N_2{}^+$, no theoretical simulation of time-dependent population transitions between the rovibronic states has been published. Therefore, by adopting the theoretical model developed for simulation of vibronic excitation of N_2^+ [22, 23, 33] and by including the rotational levels to construct complete wave functions, we can simulate time-dependent rotational excitation in N_2^+ to clarify the rotational effect in air-lasing, so that we will be able to simulate theoretically the role of the electronic, vibrational, and rotational dynamics of N_2^+ in the air lasing [39].

2.2 Theoretical Model

A nitrogen molecular ion, N_2^+ can be treated as a multi-level system composed only of several vibrational levels and their rotational levels in the respective three electronic states, $X^2\Sigma_g^+$, $A^2\Pi_u$, and $B^2\Sigma_u^+$. The time-dependent population transfer can be obtained by solving the time-dependent Schrödinger equation,

$$i\hbar\frac{\partial}{\partial t}\Psi(\boldsymbol{r},t) = H(t)\Psi(\boldsymbol{r},t), \tag{2.6}$$

where $\hbar$ is Planck's constant divided by 2π, and H is the Hamiltonian operator.

The general solution can be expressed as

$$\Psi(\mathbf{r},t) = \sum_{\alpha=\mathrm{X,A_+,A_-,B}}\ \sum_{v=0}^{V_{\max}}\sum_{K=0}^{K_{\max}}\sum_{m=-K}^{K} c_{\alpha vKm}(t)\psi_{\alpha vKm}(\mathbf{r}), \tag{2.7}$$

where $|\mathbf{r}|$ is the internuclear distance, $P_{\alpha vKm} = |c_{\alpha vKm}(t)|^2$ represents the probability of finding the system in the rotational state with quantum numbers K, m in the vth vibrational state in the electronic state α. In (2.7), α denotes one of the four lowest-energy electronic states, $X^2\Sigma_g^+$, the doubly degenerate $A^2\Pi_u$, labeled as A_+ and A_-, and $B^2\Sigma_u^+$, v is the vibrational quantum number, K is the total angular momentum in which the electron spin angular momenta are excluded, m is the projection of the total angular momentum onto the z-axis, and $\mathbf{r}$ is the internuclear separation vector. In the simulation, the maximum vibrational quantum number and the maximum rotational quantum number included in each electronic state are $V_{\max}$ and $K_{\max}$, respectively. $\psi_{\alpha vKm}(\mathbf{r})$ represents the complete molecular basis wave function of the Kth rotational state with magnetic quantum number m in the vth vibrational state in the electronic α state.

2.2.1 *Vibrational Basis*

The field-free basis wavefunction of N_2^+ is calculated by the finite difference method with the potential curves of $X^2\Sigma_g^+$, $A^2\Pi_u$ and $B^2\Sigma_u^+$ given in [30, 31]. The vibrational basis set is obtained as the eigen functions of Hamiltonian H_0 with no electronic field interaction:

$$H_0\psi_{\alpha vKm}(r) = \varepsilon_{\alpha vKm}\psi_{\alpha vKm}(r), \tag{2.8}$$

and

$$H_0 = -\frac{\hbar^2}{2\mu}\frac{\partial^2}{\partial r^2} + V_\alpha(r) + \frac{K(K+1)-k^2}{2\mu r^2}, \tag{2.9}$$

where k is the projection of the electronic orbital angular momentum on the molecular axis connecting two nitrogen nuclei. To make use of the finite difference method, we divide the space into a finite number of equally spaced nodes, separated by Δr. Thus $\psi_{\alpha vKm}(r)$ becomes a vector with elements of its value at each point over the space:

$$\psi_{\alpha vKm}(r) = \begin{pmatrix} \psi_{\alpha vKm}(r_1) \\ \psi_{\alpha vKm}(r_2) \\ \vdots \\ \psi_{\alpha vKm}(r_{\max}) \end{pmatrix}, \tag{2.10}$$

where $r_l = l \cdot \Delta r$. Since r stands for the internuclear distance of N_2^+, r can only take positive values, and is limited in the potential curve region of $(0, r_{\max})$.

The second derivative of the kinetic part in H_0 is estimated from the first derivative. Because the forward first-order derivative at $r = r_l$ is given by

$$\frac{\partial}{\partial r}\psi_{\alpha vKm}|_{r=r_l\text{forward}} = \frac{\psi_{\alpha vKm}(r_{l+1}) - \psi_{\alpha vKm}(r_l)}{\Delta r} \tag{2.11}$$

and the backward first-order derivative at $r = r_l$ is given by

$$\frac{\partial}{\partial r}\psi_{\alpha vKm}|_{r=r_l\text{backward}} = \frac{\psi_{\alpha vKm}(r_l) - \psi_{\alpha vKm}(r_{l+1})}{\Delta r}, \tag{2.12}$$

the second derivative can be expressed as

$$\begin{aligned} \frac{\partial^2}{\partial r^2}\psi_{\alpha vKm}(r)|_{r=r_l} &= \frac{1}{\Delta r}\left(\frac{\partial}{\partial r}\psi_{\alpha vKm}|_{r=r_l\text{forward}} - \frac{\partial}{\partial r}\psi_{\alpha vKm}|_{r=r_l\text{backward}}\right) \\ &= \frac{\psi_{\alpha vKm}(r_{l+1}) - 2\psi_{\alpha vKm}(r_l) + \psi_{\alpha vKm}(r_{l-1})}{\Delta r^2}. \end{aligned} \tag{2.13}$$

In this way, the differential operator H_0 can be mapped to a difference operator. By the substitution of (2.10) into (2.8) and the relationship shown in (2.13), the Hamiltonian can be expressed as

$$H_{il} = \begin{cases} -\frac{1}{2\mu}\frac{-2}{\Delta r^2} + V(r_i) + \frac{K(K+1)-k^2}{2\mu r_i{}^2} & \text{when } i = l, \\ -\frac{1}{2\mu}\frac{1}{\Delta r^2} & \text{when } i - l = \pm 1, \\ 0 & \text{when } i \neq l \text{ and } i - l \neq \pm 1. \end{cases} \tag{2.14}$$

Solutions of the eigenfunction above give the eigenvalues as $\{\varepsilon_{\alpha v K m}\}$, corresponding to the rovibrational energies, and the eigenvectors as $\{\psi_{\alpha v K m}(r)\}$, corresponding to the rovibrational wave function of the Kth rotational level in the νth vibrational state.

2.2.2 Rotational Basis

The standard rotational basis functions are spherical harmonics [40]. We adopt here the rotational basis wave function $|K, m, k\rangle$ defined as [41]

$$
\begin{aligned}
|K, m, k\rangle &= \frac{1}{\sqrt{2}}\sqrt{(K+m)!(K-m)!(K+k)!(K-k)!(2K+1)} \\
&\times \sum_{\sigma}(-1)^{\sigma}\frac{[\cos(\theta/2)]^{2K+k-m-2\sigma}[-\sin(\theta/2)]^{m-k+2\sigma}}{\sigma!(K-m-\sigma)!(m-k+\sigma)!(K+k-\sigma)!}e^{im\phi},
\end{aligned}
\tag{2.15}
$$

where θ is the polar angle, ϕ is the azimuth angle, and k is the projection of the electronic orbital angular momentum on the molecular axis. When $k = 0$, $|K, m, k\rangle$ becomes spherical harmonics. We have $k = 0$ for $X^2\Sigma_g{}^+$ and $B^2\Sigma_u{}^+$, while $k = \pm 1$ for $A^2\Pi_u$. Note that the wave functions $|K, m, +1\rangle$ and $|K, m, -1\rangle$ are not symmetric with respect to the transformation $\theta \to \pi - \theta$, but by the subtraction and addition, two symmetric basis sets and $|K, m\rangle^{A+}$ and $|K, m\rangle^{A-}$ for the degenerate $A^2\Pi_u$ state can be constructed as shown in (2.16). We label the corresponding rotational basis wave function on different electronic state as $|K, m\rangle^{\alpha}$, defined as

$$
|K, m\rangle^{\alpha} = \begin{cases} |K, m, 0\rangle & \text{when } \alpha = \mathrm{X}, \mathrm{B}, \\ (|K, m, 1\rangle + |K, m, -1\rangle)/2 & \text{when } \alpha = \mathrm{A}_+, \\ (|K, m, 1\rangle - |K, m, -1\rangle)/2 & \text{when } \alpha = \mathrm{A}_-. \end{cases}
\tag{2.16}
$$

2.2.3 Numerical Solution of Time-Dependent Schrödinger Equation

By substituting (2.7) into (2.6), multiplying with $\psi^*_{\beta v' K' m'}(r)$ from left, and integrating over **r**, we obtain

$$
i\hbar\frac{d}{dt}c_{\beta v' K' m'}(t) = \sum_{\alpha=\mathrm{X},\mathrm{A}_+,\mathrm{A}_-,\mathrm{B}}\sum_{v=0}^{V_{\max}}\sum_{K=0}^{K_{\max}}\sum_{m=-K}^{K} c_{\alpha v K m}(t) H_{\beta v' K' m' \alpha v K m}(t),
\tag{2.17}
$$

where

$$H_{\beta v' K' m' \alpha v K m}(t) = \int_0^{\infty}\int_0^{\pi}\int_0^{2\pi} \psi^*_{\beta v' K' m'}(\mathbf{r}) H(t) \psi_{\alpha v K m}(\mathbf{r}) dr \sin\theta d\theta d\phi. \tag{2.18}$$

Under an intense laser pulse, the total Hamiltonian H of the system interacting with the laser field becomes

$$H(t) = -\frac{\hbar^2}{2\mu}\nabla^2 + V + H_1(t) = H_0 + H_1(t), \tag{2.19}$$

where μ is the reduced mass of $N_2{}^+$, V is the interatomic potential energy, and H_1 stands for the interaction with the laser field. Therefore, $H_{\beta v' K' m' \alpha v K m}(t)$ becomes

$$H_{\beta v' K' m' \alpha v K m}(t) = (H_0)_{\beta v' K' m' \alpha v K m} + (H_1)_{\beta v' K' m' \alpha v K m}(t), \tag{2.20}$$

where

$$\begin{aligned}(H_0)_{\beta v' K' m' \alpha v K m} &= \langle \psi_{\beta v' K' m'} | H_0 | \psi_{\alpha v K m} \rangle \\ &= \varepsilon_{\alpha v K m} \delta_{\alpha\beta} \delta_{v' v} \delta_{K' K} \delta_{m' m},\end{aligned} \tag{2.21}$$

and

$$\begin{aligned}(H_1)_{\beta v' \alpha v}(t) &= \langle \psi_{\beta v' K' m'} | H_1(t) | \psi_{\alpha v K m} \rangle \\ &= \langle \psi_{\beta v' K' m'} | D_{\alpha\beta} F_\theta^{\alpha\beta} | \psi_{\alpha v K m} \rangle \mathcal{E}(t).\end{aligned} \tag{2.22}$$

In (2.22), representing the coupling between the rotational levels via the transition dipole moment $D_{\alpha\beta}$, $\mathcal{E}(t)$ stands for the laser field strength at time t defined as

$$\mathcal{E}(t) = \mathcal{E}_0 f(t) \cos(\omega t), \tag{2.23}$$

where $f(t)$ is an envelope function, ω is the frequency of the laser pulse. $F_\theta^{\alpha\beta}$ is the angular factor representing the polar-angle dependent coupling between the X, A and B states defined as

$$F_\theta^{\alpha\beta} = \cos(\theta)(\delta_{B\alpha}\delta_{X\beta} + \delta_{X\alpha}\delta_{B\beta}) + \sin(\theta)(\delta_{A\alpha}\delta_{X\beta} + \delta_{X\alpha}\delta_{A\beta}). \tag{2.24}$$

The envelope function is defined as

$$f(t) = e^{-t^2/2\sigma_0^2}, \tag{2.25}$$

where σ is chosen according to the intensity half-width of the laser pulse adopted from the experimental conditions and the simulation is conducted from $t = 0$, resulting in a half laser pulse with a sudden turn-on behavior.

2.2.4 Quasi-stationary Floquet Theory

While Floquet theory is applicable when the light field amplitude is constant, the quasi-stationary Floquet theory is applicable when the amplitude of a light field varies slowly. In the case of N_2^+, as long as the generated ions experience the latter half of the laser pulse whose amplitude vary slowly the quasi-stationary Floquet theory can be used. The wave function of the N_2^+ system can be expressed as a linear combination of Floquet states,

$$\Psi(r,t) = \sum_q k_q(t) e^{-i\varepsilon_q^F(t)t} \phi_q(r,t), \tag{2.26}$$

where $\varepsilon_q^F(t)$ is the quasi-energy and $\phi_q(r,t)$ is the wave function of the qth Floquet state, which is constructed from the field-free basis derived in Sect. 2.2.1.

$$\phi_q(r,t) = \sum_{n=-\infty}^{\infty} \sum_{\alpha=X,A,B} \sum_{v=0}^{v_{max}} \phi_{\alpha v,q}^{(n)}(t) \psi_{\alpha v}(r) e^{-in\omega t}. \tag{2.27}$$

Because the rotational motion does not affect the mechanism of the population inversion between the vibrational levels in the electronic X and B state, the k and m can be omitted so that the field-free vibrational basis becomes $\psi_{\alpha v}(r)$. The time-dependent coefficient $k_q(t)$ is the amplitude of the qth Floquet state. The time-dependence of $k_q(t)$ can be expressed as

$$\frac{dk_q(t)}{dt} = -\mathcal{E}_0 \frac{df(t)}{dt} \sum_p \left\langle \Psi_q^{\mathcal{E}_0 f(t)} \middle| \frac{\partial}{\partial \mathcal{E}_0} \Psi_p^{\mathcal{E}_0 f(t)} \right\rangle k_p(t), \tag{2.28}$$

where

$$\Psi_q^{\mathcal{E}_0 f(t)}(r,t) = e^{-i\varepsilon_q^F(t)t} \phi_q(r,t) \tag{2.29}$$

is a Floquet state obtained at time t at a given value of $\mathcal{E}_0 f(t)$. If the coupling between the field-free states is off-resonance, and the envelope of the pulse changes slowly, that is, if the coupling is in the adiabatic limit, $\frac{df(t)}{dt} = 0$, $k_q(t)$ becomes a constant independent of time, and the quasi-stationary Floquet theory can be applied. However, if the coupling between the field-free states is on-resonance, non-adiabatic population transfer occurs between the Floquet states, and consequently, the quasi-stationary Floquet theory cannot be used [42].

When the pulsed light field having an envelope function vanishing gradually in time, the Floquet states approach adiabatically to the field-free states. If the qth Floquet state is transformed into the vth vibrational state in the α electronic state, that is, when

$$\lim_{t\to\infty} \phi_q(t) = \psi_{\alpha v}, \tag{2.30}$$

is fulfilled, the qth Floquet state is labeled as the $\alpha(v)$ Floquet state.

2.3 Population Inversion in Aligned N_2^+

In most of the analysis of the results of the experiments in which the 391 nm coherent emission from N_2^+ was recorded, rotational motion of N_2^+ has been neglected because the rotational period ($\sim$ 8 ps) [43] is much longer than the laser pulse duration ($\sim$ 40 fs) [6]. Therefore, it can be assumed that N_2^+ interacts with the laser field as if N_2^+ is fixed in the space. As mentioned in Sect. 2.2.3, the polar angle θ, can be regarded as the alignment angle between the molecular axis and polarization direction of the laser pulse.

In the simulation of the population transfer among the vibronic levels of N_2^+ interacting with a laser pulse, N_2^+ is assumed to be created by the tunneling ionization of N_2 at the moment when the electric field of the near-IR pulse takes the maximum amplitude during on optical cycle, so that the excitation is calculated from $t = 0$ with an envelope function defined in (2.25).

The total wave function of the resultant N_2^+ is expressed using the field-free vibrational basis set $\{\psi_{\alpha v}(r)\}$ obtained by the same method in Sect. 2.2.1 without the rotational motion as

$$\Psi(r,t) = \sum_{\alpha=\mathrm{X,A,B}} \sum_{v=0}^{V_{max}} c_{\alpha v}(t)\psi_{\alpha v}(r). \tag{2.31}$$

Upon the creation of N_2^+ at $t = 0$, the vibrational wave packet of N_2^+ is assumed to be the same as that in the vibrational ground state of the electronic ground state $X^2\Sigma_g{}^+$ of neutral N_2 according to the Franck Condon principle as

$$c_{\alpha v}(t=0) = \left\langle \psi_{\alpha,v} \middle| \psi_{v=0}^{N_2} \right\rangle \delta_{\alpha X}, \tag{2.32}$$

where α = X, A, and B.

2.3.1 Time-Dependent Population Transition

When N_2^+ is created by the laser pulse, the initial states are calculated by the projection of the wave function in the vibrational ground state of neutral N_2 onto the three electronic states in N_2^+. By solving the time-dependent Schrödinger equation, with the total wavefunction given by (2.31) and the experimental conditions reported in [6], the post-ionization coupled-state dynamics is obtained as shown in Fig. 2.3. The simulation is done by employing a laser pulse with a full-width at half-maximum (FWHM) of 20 fs, the field intensity of 2×10^{14} Wcm^{-2}, and the alignment angle of $\theta = 45°$. In Fig. 2.3, populations in the respective electronic states oscillate fast and become constants after the laser pulse vanishes, that is, after the interaction with the

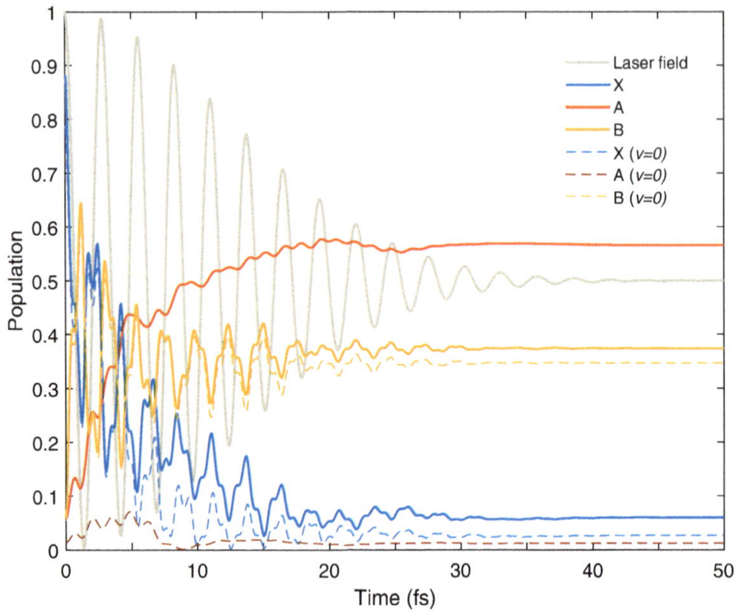

Fig. 2.3 Time-dependent populations of the three electronic states simulated with the same laser pulse conditions as in [6]. Thick lines represent the total population in the respective vibrational states and thin lines correspond to the populations in the vibrational ground states. The laser electric field is plotted in grey in the background on an arbitrary scale

laser pulse, the population inversion between the vibrational ground state of $B^2\Sigma_u{}^+$ and that of $X^2\Sigma_g{}^+$ is achieved.

As discussed in [32], the amount of population transfered from the ground level to the excited level in a two-level system at off resonance depends on the field intensity and the coupling strength, combined in a parameter

$$a = -\frac{\mu_{XB}\mathcal{E}_0}{2\hbar\omega}, \tag{2.33}$$

$$\mu_{XB} = \langle\psi_B|\, D_{BX}\, |\psi_X\rangle\,, \tag{2.34}$$

representing the coupling strength between $B^2\Sigma_u{}^+$ and $X^2\Sigma_g{}^+$. When the N-N molecular axis of $N_2{}^+$ is parallel to the polarization direction of the laser field, there will be no population transfer from $X^2\Sigma_g{}^+$ to $A^2\Pi_u$, and therefore, only two electronic states are involved in the post-ionization excitation. The population transferred to $B^2\Sigma_u{}^+$ is plotted in Fig. 2.4 as a function of a, with the results obtained for the two-level system [32]. As shown in Fig. 2.4, the trends that the final population in the excited state increases when the laser field intensity increases are common in both models, which implies that N_2^+ can roughly be treated as a two-level system.

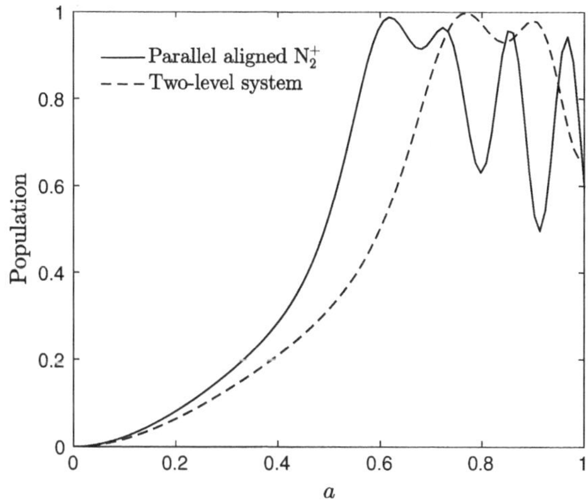

Fig. 2.4 Final population transferred to the excited state as a function of the coupling strength. Solid line represents the population transferred to $B^2\Sigma_u{}^+$, and the dashed line represents the population transferred to the excited state. The laser field conditions are the same as those in Fig. 2.3

2.3.2 *Floquet Interpretation*

In order to interpret the mechanism of the efficient population inversion achieved in $N_2{}^+$, we solve the time-dependent Schrödinger equations numerically and obtain the time-dependent populations in the Floquet states of $N_2{}^+$.

In the calculation, four vibrational states are included in the respective electronic states and the dressed photon number is taken up $m = 6$. In Fig. 2.1a, the populations in the seven most populated Floquet states are plotted. In the calculation, the angle θ between the N–N molecular axis and the polarization direction of the laser field was set to be 45°.

As seen in Fig. 2.5, the $B(v = 0)$ Floquet state has 15% population when the field is suddenly turned on. The population in the $B(v = 0)$ Floquet is almost constant during the interaction with the half laser pulse after N_2^+ is generated by the ionization process. The constant population suggests the adiabatic property of the $B(v = 0)$ Floquet state, which eventually becomes the vibrational ground state of the B state. At the same time, in the state is also transferred to the A Floquet states through the non-adiabatic Floquet coupling between the X and A states at one-photon resonance. Finally, when the pulse vanishes, the population in the field-free $X(v = 0)$ state is smaller than that of the $B(v = 0)$ state, resulting in the population inversion between the X and B states.

In order to analyze the initial population in the $B(v = 0)$ state, we decompose the $B(v = 0)$ Floquet state in the time domain. As shown in Fig. 2.6, the components from $X(v = 0)$ and $X(v = 1)$, both of which are dressed by 0 photon, have significant contribution to the $B(v = 0)$ Floquet state at $t = 0$, which ensures a creation of a certain amount of population in the $B(v = 0)$ Floquet state when the pulse is suddenly turned on. As time evolves, the contribution from $X^2\Sigma_g{}^+$ in this Floquet

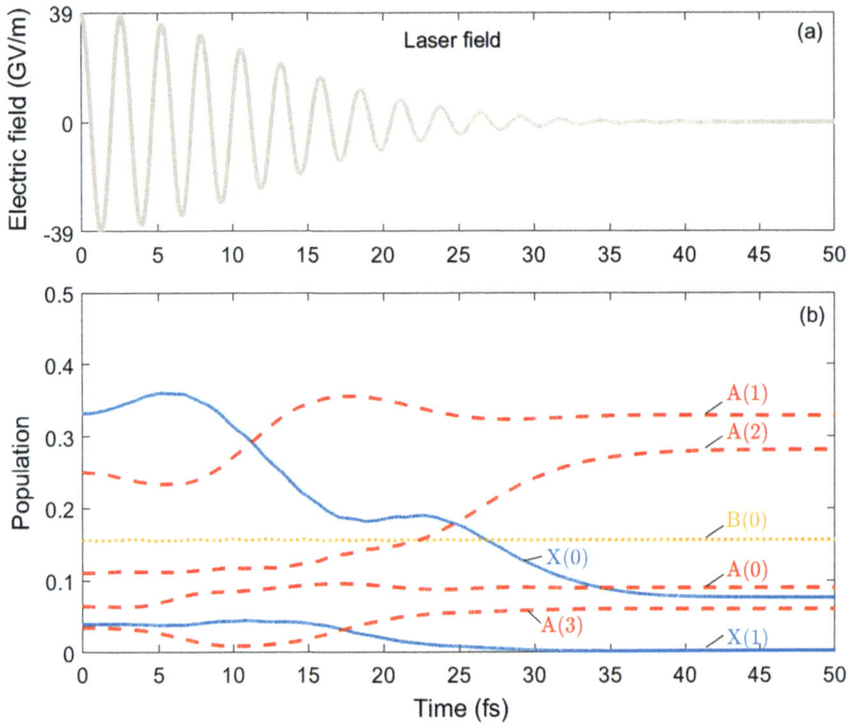

Fig. 2.5 **a** Time-dependent electric field of the laser pulse employed in the simulation. **b** Time-dependent populations in the Floquet states of $N_2{}^+$ at the field intensity of 2×10^{14} Wcm^{-2}. The label "$\alpha(n)$" stands for "the $\alpha(v = n)$ Floquet state". The angle θ between the molecular axis of $N_2{}^+$ and the laser polarization direction is set to be $\theta = 45°$. Only the seven most populated states are shown here. The initial populations in the $X^2\Sigma_g{}^+$, $A^2\Pi_u$, and $B^2\Sigma_u{}^+$ states are $\sum_{v=0}^{V\max} |c_{X,v}(t = 0)|^2 = 1$, $|c_{A,v}(t = 0)|^2 = 0$ and $|c_{B,v}(t = 0)|^2 = 0$, respectively

state gradually decreases to 0 and the contribution from $B^2\Sigma_u{}^+$, dressed by 1 photon, gradually increases from around 0.9 to 1.0, and finally this Floquet state becomes the vibrational ground state of $B^2\Sigma_u{}^+$.

2.3.3 Population Inversion Assisted by $A^2\Pi_u$ *State*

Because the coupling between $B^2\Sigma_u{}^+$ and $X^2\Sigma_g{}^+$ is orthogonal to that between $A^2\Pi_u$ and $X^2\Sigma_g{}^+$, by varying the value of alignment angle θ, the relative contribution of these transitions can be adjusted. When θ increases from 0° to 90°, the direction of the N-N molecular axis of $N_2{}^+$ changes from parallel to perpendicular with respect to the polarization direction of the laser field.

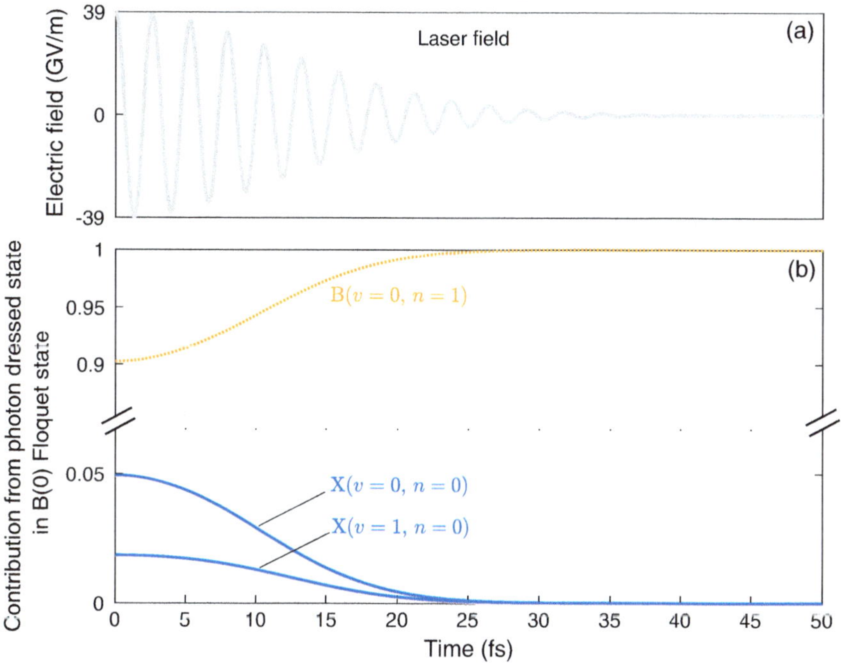

Fig. 2.6 **a** Time-dependent electric field of the laser pulse employed in the simulation. **b** Time-dependent contributions from the photon-dressed components of the $B(v = 0)$ Floquet state

The difference in the populations in $B^2\Sigma_u{}^+(v = 0)$ and $X^2\Sigma_g{}^+(v = 0)$ is plotted in Fig. 2.7 as a function of the alignment angle and the field strength parameter a, which is defined as

$$a = -\frac{\mu_{BX}\mathcal{E}_0}{2\hbar\omega}, \tag{2.35}$$

where the transition dipole matrix element μ is given by

$$\mu_{BX} = \langle\psi_{X0}|\, D_{XB}\, |\psi_{B0}\rangle \,. \tag{2.36}$$

As shown clearly in Fig. 2.7, when the field intensity is 2×10^{14} Wcm^{-2}, corresponding to $a = 0.4$, the population inversion can be achieved between $B^2\Sigma_u{}^+$ and $X^2\Sigma_g{}^+$ only when θ is in the range of $22° < \theta < 62°$. The angle range becomes narrower if the laser field intensity becomes smaller than $a = 0.4$ (2×10^{14} Wcm^{-2}). At a specific alignment angle, for example, at $\theta = 70°$, the population difference between $B^2\Sigma_u{}^+$ and $X^2\Sigma_g{}^+$ first increases and then decreases when $a > 0.35$.

Even though $A^2\Pi_u$ state does not contribute to the population transfer between $X^2\Sigma_g{}^+$ and $B^2\Sigma_u{}^+$, it promotes the population inversion between $X^2\Sigma_g{}^+$ and $B^2\Sigma_u{}^+$ by depleting the population from the $X^2\Sigma_g{}^+$ state through the transition from the $X^2\Sigma_g{}^+$ state to the $A^2\Pi_u$ state.

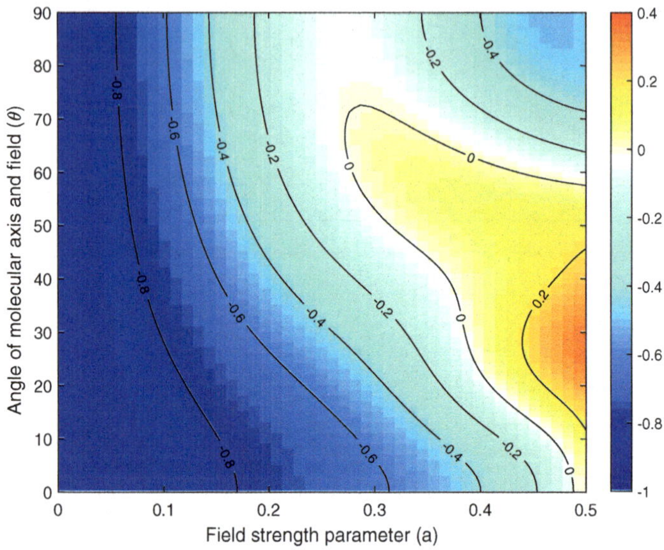

Fig. 2.7 Contour plot of the difference in the final population in the B(v = 0) state and that in the X(v = 0) state as a function of the laser field strength in the range between $0 < a < 0.5$ and the angle θ between the N-N molecular axis and the polarization direction of the laser field. When the population in the B(v = 0) state is larger than that in the X(v = 0) state, the population difference is represented as a positive value, that is, $\Delta P = P(\mathrm{B}(v=0)) - P(\mathrm{X}(v=0))$, where P stands for a population in the state in the parentheses. The initial populations in the $\mathrm{X}^2\Sigma_g{}^+$, $\mathrm{A}^2\Pi_u$, and $\mathrm{B}^2\Sigma_u{}^+$ states are $\sum_{v=0}^{Vmax} |c_{X,v}(t=0)|^2 = 1$, $|c_{A,v}(t=0)|^2 = 0$ and $|c_{B,v}(t=0)|^2 = 0$, respectively

2.4 Rotational Excitation in $N_2{}^+$

By including the electronic, vibrational and rotational degrees of freedom, the population transfer among the rotational levels of $N_2{}^+$ interacting with a sudden turn-on pulse can be obtained. The calculation starts from each pure rotational initial state $c_{X0Km}(t=0) = 1$ by solving the Schrödinger equation with a sufficient small time step at $T = 300$ K. The population in the rotational level $|\alpha v K' m\rangle$ at time t calculated from the initial state $c_{X0Km}(t=0) = 1$ is labeled as $|c_{\alpha v K' m}^{X0Km}(t)|^2$, so that the thermally averaged population in the rotational level $|\alpha v K' m\rangle$ is expressed by using the Boltzmann distribution as

$$|c_{\alpha v K'}(t)|^2 = \sum_{K=0}^{K_{\max}} \sum_{m=-K}^{K} |c_{\alpha v K' m}^{X0Km}(t)|^2 \frac{g_K e^{-\frac{B_{\alpha,K} K(K+1)}{k_B T}}}{\zeta}, \tag{2.37}$$

where $B_{\alpha,K}$ is the rotational constant, k_B is the Boltzmann constant, T is the temperature, and the nuclear spin factor is $g_K = 2$ when K is even and $g_K = 1$ when K is odd. The normalization factor ζ is given by

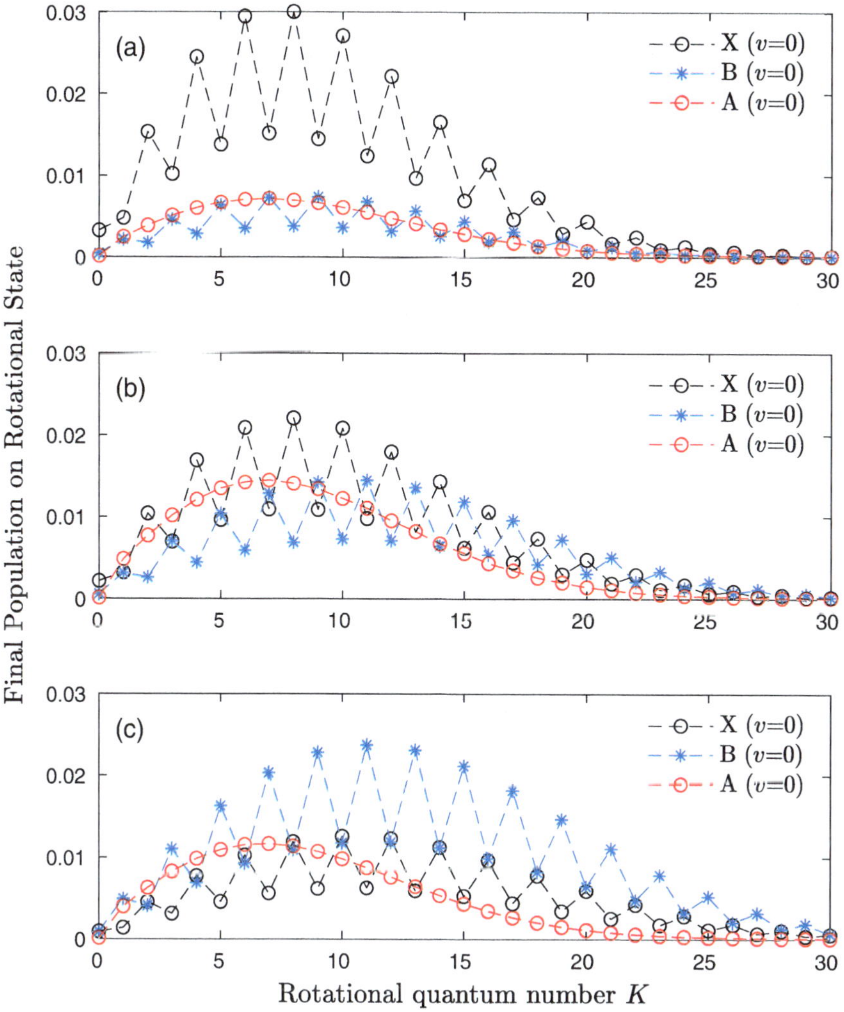

Fig. 2.8 Final population distribution in the rotational levels of N_2^+ obtained after the interaction with an intense 800 nm laser pulse with the field intensities of **a** 2×10^{14} Wcm^{-2}, **b** 4×10^{14} Wcm^{-2} and **c** 6×10^{14} Wcm^{-2}. The label "$\alpha(v = n)$" stands for the nth vibrational state of the α state. The initial populations in the $X^2\Sigma_g^+$, $A^2\Pi_u$, and $B^2\Sigma_u^+$ states are $\sum_{v=0}^{Vmax} |c_{X,v}(t=0)|^2 = 1$, $|c_{A,v}(t=0)|^2 = 0$ and $|c_{B,v}(t=0)|^2 = 0$, respectively

$$\zeta = \sum_{K=0}^{K_{\max}} g_K(2K+1)e^{-\frac{B_{\alpha,K}K(K+1)}{k_BT}}. \tag{2.38}$$

The $K_{\max}$ in (2.37) is set to be 40.

Because N_2^+ is estimated to be prepared in the electronic ground $X^2\Sigma_g^+$ state with 90% probability according to molecular Ammosov-Delone-Krainov (MO-ADK) theory [44] and because the Franck-Condon factor for the ionization from the vibrational ground state of N_2 to the vibrational ground state of N_2^+ is 0.9, N_2^+ ion generated by the intense laser pulse is assumed to be prepared 100% in the vibrational ground level in the $X^2\Sigma_g^+$ state (labeled as X ($v = 0$)) at $t = 0$.

As shown in Fig. 2.8a, population inversion is not achieved between $X^2\Sigma_g^+$ and $B^2\Sigma_u^+$ when the field intensity is 2×10^{14} Wcm^{-2}. On the other hand, as shown in Fig. 2.8c, when the field intensity becomes sufficiently large, the population inversion between the rotational states of $X^2\Sigma_g^+$ and $B^2\Sigma_u^+$ is achieved in the wide range of the K quantum number so that the net population inversion is achieved between X ($v = 0$) and B ($v = 0$). At the medium field intensity 4×10^{14} Wcm^{-2}, it can be seen in Fig. 2.8b that B ($v = 0, K$) is more populated than X ($v = 0, K - 1$) when $K \geq 15$ and that B ($v = 0, K$) is more populated than X ($v = 0, K + 1$) and X ($v = 0, K - 1$) when $K \geq 21$, showing that the lasing at the P-branch emission is realized when $K \geq 15$ and that the lasing at the R-branch emission can also be realized when $K \geq 21$, which is consistent with the experimental results reported in [28].

2.5 Summary

Our recent theoretical approaches to exploring the mechanism of population inversion in N_2^+, resulting in the generation of the coherent emission at 391 nm called air-lasing, have been reviewed. Because the ionization preferentially proceeds at the peak laser field intensity within one optical cycle, N_2^+ created in the laser field starts interacting with the intense laser field immediately after its birth and is excited electronically by the laser field. This experimental situation can be simulated theoretically as the optical excitation of N_2^+ by a sudden turn-on pulse [6].

We demonstrated by quasi-stationary Floquet theory [32] that an efficient population transfer occurs in a two-level system when it interacts with a sudden turn-on pulse even when the carrier frequency is off-resonance with the energy separation of the two levels. We applied this sudden turn-on model to N_2^+ interacting with an intense near-IR laser field and confirmed that the population in the light-dressed $B^2\Sigma_u^+$ is adiabatically transformed to the field free B ($v = 0$) state when the pulsed laser field vanishes [33]. We also confirmed on the basis of the sudden turn-on model that the population inversion between B ($v = 0$) and X ($v = 0$) can be achieved rotationally even when the net population inversion is not achieved between B ($v = 0$) and X ($v = 0$). Furthermore, the resonant transition between $A^2\Pi_u$ and $X^2\Sigma_g^+$ contributes to the decrease in the final population in the $X^2\Sigma_g^+$ and to the modification of the rotational distributions in the vibronic states involved in the air-lasing. The sudden turn-on model that we have been developing in the past several years to investigate the population transfer among the rovibronic levels of N_2^+ can be applied to the investigation of time-dependent population transfer processes of any atomic and molecular ions that are created suddenly within an intense laser field.

Acknowledgements This research was supported by JSPS KAKENHI grants no. JP15K17805, no. JP24245003, and no. JP15H05696.

References

1. V. Kocharovsky, S. Cameron, K. Lehmann, R. Lucht, R. Miles, Y. Rostovtsev, W. Warren, G.R. Welch, M.O. Scully, Gain-swept superradiance applied to the stand-off detection of trace impurities in the atmosphere. Proc. Nat. Acad. Sci. **102**, 7806 (2005)
2. P.R. Hemmer, R.B. Miles, P. Polynkin, T. Siebert, A.V. Sokolov, P. Sprangle, M.O. Scully, Standoff spectroscopy via remote generation of a backward-propagating laser beam. Proc. Natl. Acad. Sci. U S A **108**, 3130 (2011) 21297033 [pmid]
3. Q. Luo, W. Liu, S. Chin, Lasing action in air induced by ultra-fast laser filamentation. Appl. Phys. B **76**, 337 (2003)
4. A. Dogariu, J.B. Michael, M.O. Scully, R.B. Miles, High-gain backward lasing in air. Science **331**, 442 (2011)
5. J. Yao, B. Zeng, H. Xu, G. Li, W. Chu, J. Ni, H. Zhang, S.L. Chin, Y. Cheng, Z. Xu, High-brightness switchable multiwavelength remote laser in air. Phys. Rev. A **84**, 051802 (2011)
6. H. Xu, E. Lötstedt, A. Iwasaki, K. Yamanouchi, Sub-10-fs population inversion in N_2^+ in air lasing through multiple state coupling. Nat. Commun. **6**, 8347 (2015)
7. A. Zhang, Q. Liang, M. Lei, L. Yuan, Y. Liu, Z. Fan, X. Zhang, S. Zhuang, C. Wu, Q. Gong, H. Jiang, Coherent modulation of superradiance from nitrogen ions pumped with femtosecond pulses. Opt. Express **27**, 12638 (2019)
8. J. Chen, J. Yao, H. Zhang, Z. Liu, B. Xu, W. Chu, L. Qiao, Z. Wang, J. Fatome, O. Faucher, C. Wu, Y. Cheng, Electronic-coherence-mediated molecular nitrogen-ion lasing in a strong laser field. Phys. Rev. A **100**, 031402 (2019)
9. H. Xie, Q. Zhang, G. Li, X. Wang, L. Wang, Z. Chen, H. Lei, Z. Zhao, Vibrational population transfer between electronic states of N_2^+ in polarization modulated intense laser fields. Phys. Rev. A **100**, 053419 (2019)
10. G. Li, C. Jing, B. Zeng, H. Xie, J. Yao, W. Chu, J. Ni, H. Zhang, H. Xu, Y. Cheng, Z. Xu, Signature of superradiance from a nitrogen-gas plasma channel produced by strong-field ionization. Phys. Rev. A **89**, 033833 (2014)
11. Y. Liu, P. Ding, G. Lambert, A. Houard, V. Tikhonchuk, A. Mysyrowicz, Recollision-induced superradiance of ionized nitrogen molecules. Phys. Rev. Lett. **115**, 133203 (2015)
12. X. Zhong, Z. Miao, L. Zhang, Q. Liang, M. Lei, H. Jiang, Y. Liu, Q. Gong, C. Wu, Vibrational and electronic excitation of ionized nitrogen molecules in intense laser fields. Phys. Rev. A **96**, 043422 (2017)
13. X. Zhong, Z. Miao, L. Zhang, H. Jiang, Y. Liu, Q. Gong, C. Wu, Optimizing the 391-nm lasing intensity from ionized nitrogen molecules in 800-nm femtosecond laser fields. Phys. Rev. A **97**, 033409 (2018)
14. J. Ni, W. Chu, C. Jing, H. Zhang, B. Zeng, J. Yao, G. Li, H. Xie, C. Zhang, H. Xu, S.-L. Chin, Y. Cheng, Z. Xu, Identification of the physical mechanism of generation of coherent N_2^+ emissions in air by femtosecond laser excitation. Opt. Express **21**, 8746 (2013)
15. J. Yao, S. Jiang, W. Chu, B. Zeng, C. Wu, R. Lu, Z. Li, H. Xie, G. Li, C. Yu, Z. Wang, H. Jiang, Q. Gong, Y. Cheng, Population redistribution among multiple electronic states of molecular nitrogen ions in strong laser fields. Phys. Rev. Lett. **116**, 143007 (2016)
16. V.T. Tikhonchuk, J.-F. Tremblay-Bugeaud, Y. Liu, A. Houard, A. Mysyrowicz, Excitation of nitrogen molecular ions in a strong laser field by electron recollisions. Euro. Phys. J. D **71**, 292 (2017)
17. M. Britton, P. Laferrière, D.H. Ko, Z. Li, F. Kong, G. Brown, A. Naumov, C. Zhang, L. Arissian, P.B. Corkum, Testing the role of recollision in N_2^+ air lasing. Phys. Rev. Lett. **120**, 133208 (2018)

18. H. Li, Q. Song, J. Yao, Z. Liu, J. Chen, B. Xu, K. Lin, J. Qiang, B. He, H. Xu, Y. Cheng, H. Zeng, J. Wu, Air lasing from singly ionized N_2 driven by bicircular two-color fields. Phys. Rev. A **99**, 053413 (2019)
19. A. Zhang, M. Lei, J. Gao, C. Wu, Q. Gong, H. Jiang, Subfemtosecond-resolved modulation of superfluorescence from ionized nitrogen molecules by 800-nm femtosecond laser pulses. Opt. Express **27**, 14922 (2019)
20. Y. Wan, B. Xu, J. Yao, J. Chen, Z. Liu, F. Zhang, W. Chu, Y. Cheng, Polarization ellipticity dependence of $N_2{}^+$ air lasing: the role of coupling between the ground state and a photo-excited intermediate state. J. Opt. Soc. Am. B **36**, G57 (2019)
21. A. Mysyrowicz, R. Danylo, A. Houard, V. Tikhonchuk, X. Zhang, Z. Fan, Q. Liang, S. Zhuang, L. Yuan, Y. Liu, Lasing without population inversion in $N_2{}^+$. APL Photon. **4**, 110807 (2019)
22. H. Xu, E. Lötstedt, T. Ando, A. Iwasaki, K. Yamanouchi, Alignment-dependent population inversion in $N_2{}^+$ in intense few-cycle laser fields. Phys. Rev. A **96**, 041401 (2017)
23. H. Li, M. Hou, H. Zang, Y. Fu, E. Lötstedt, T. Ando, A. Iwasaki, K. Yamanouchi, H. Xu, Significant enhancement of $N_2{}^+$ lasing by polarization-modulated ultrashort laser pulses. Phys. Rev. Lett. **122**, 013202 (2019)
24. H. Zhang, C. Jing, J. Yao, G. Li, B. Zeng, W. Chu, J. Ni, X. Hongqiang, H. Xu, S. Leang Chin, K. Yamanouchi, H. Sun, Z. Xu, Rotational coherence encoded in an "air-laser" spectrum of nitrogen molecular ions in an intense laser field. Phys. Rev. X **3** (2013)
25. B. Zeng, W. Chu, G. Li, J. Yao, H. Zhang, J. Ni, C. Jing, H. Xie, Y. Cheng, Real-time observation of dynamics in rotational molecular wave packets by use of air-laser spectroscopy. Phys. Rev. A **89**, 042508 (2014)
26. H. Xie, B. Zeng, G. Li, W. Chu, H. Zhang, C. Jing, J. Yao, J. Ni, Z. Wang, Z. Li, Y. Cheng, Coupling of $N_2{}^+$ rotational states in an air laser from tunnel-ionized nitrogen molecules. Phys. Rev. A **90**, 042504 (2014)
27. M. Lei, C. Wu, A. Zhang, Q. Gong, H. Jiang, Population inversion in the rotational levels of the superradiant $N_2{}^+$ pumped by femtosecond laser pulses. Opt. Express **25**, 4535 (2017)
28. A. Azarm, P. Corkum, P. Polynkin, Optical gain in rotationally excited nitrogen molecular ions. Phys. Rev. A **96**, 051401 (2017)
29. M. Britton, M. Lytova, P. Laferrière, P. Peng, F. Morales, D.H. Ko, M. Richter, P. Polynkin, D.M. Villeneuve, C. Zhang, M. Ivanov, M. Spanner, L. Arissian, P.B. Corkum, Short- and long-term gain dynamics in $N_2{}^+$ air lasing. Phys. Rev. A **100**, 013406 (2019)
30. S.R. Langhoff, C.W. Bauschlicher, H. Partridge, Theoretical study of the $N_2{}^+$ Meinel system. J. Chem. Phys. **87**, 4716 (1987)
31. S.R. Langhoff, C.W. Bauschlicher Jr., Theoretical study of the first and second negative systems of $N_2{}^+$. J. Chem. Phys. **88**, 329 (1988)
32. Y. Zhang, E. Lötstedt, K. Yamanouchi, Population inversion in a strongly driven two-level system at far-off resonance. J. Phys. B: At. Mol. Opt. Phys. **50**, 185603 (2017)
33. Y. Zhang, E. Lötstedt, K. Yamanouchi, Mechanism of population inversion in laser-driven $N_2{}^+$. J. Phys. B: At. Mol. Opt. Phys. **52**, 055401 (2019)
34. W. Zheng, Z. Miao, L. Zhang, Y. Wang, C. Dai, A. Zhang, H. Jiang, Q. Gong, C. Wu, Enhanced coherent emission from ionized nitrogen molecules by femtosecond laser pulses. J. Phys. Chem. Lett. **6598** (2019)
35. S. Tamar, Rotational excitation and molecular alignment in intense laser fields. J. Chem. Phys. **103**, 7887 (1995)
36. L. Matsuoka, E. Segawa, Localization in rotational excitation of diatomic molecules induced by a train of optical pulses. Interdisc. Inf. Sci. **23**, 51 (2017)
37. A. Maan, A. Tyagi, V. Prasad, Rotational excitation of diatomic molecule: time dependent study. Acta Phys. Pol., A **133**, 1266 (2018)
38. T. Szidarovszky, M. Jono, K. Yamanouchi, LIMAO: cross-platform software for simulating laser-induced alignment and orientation dynamics of linear-, symmetric- and asymmetric tops. Comput. Phys. Commun. **228**, 219 (2018)
39. T. Ando, E. Lötstedt, A. Iwasaki, H. Li, Y. Fu, S. Wang, H. Xu, K. Yamanouchi, Rotational, vibrational, and electronic modulations in $N_2{}^+$ lasing at 391 nm: evidence of coherent $B^2\Sigma_u^+ - X^2\Sigma_g^+ - A^2\Pi_u$ Coupling. Phys. Rev. Lett. **123**, 203201 (2019)

40. M.E. Rose, *Elementary Theory of Angular Momentum* (Wiley, New York, NY, 1957)
41. J.M. Brown, A. Carrington, *Rotational Spectroscopy of Diatomic Molecules, Cambridge Molecular Science* (Cambridge University Press, Cambridge, 2003)
42. S.-I. Chu, D.A. Telnov, Beyond the Floquet theorem: generalized Floquet formalisms and quasienergy methods for atomic and molecular multiphoton processes in intense laser fields. Phys. Rep. **390**, 1 (2004)
43. X. Li, J. Paldus, Full potential energy curve for N_2 by the induced multireference coupled-cluster method. J. Chem. Phys. **129**, 054104 (2008)
44. S.-F. Zhao, C. Jin, A.-T. Le, T.F. Jiang, C.D. Lin, Determination of structure parameters in strong-field tunneling ionization theory of molecules. Phys. Rev. A **81**, 033423 (2010)

Chapter 3
Bichromatic Control of Free Electron Wave Packets

Stefanie Kerbstadt, Kevin Eickhoff, Tim Bayer, and Matthias Wollenhaupt

Abstract We report on recent developments in coherent control of ultrafast photoionization dynamics with ultrashort polarization-shaped bichromatic pulse sequences. In the experiments, a novel pulse shaping scheme for the generation of **Bi**chromatic **C**arrier-**E**nvelope phase-stable **P**olarization-tailored **S**upercontinuum (BiCEPS) fields is combined with high-resolution photoelectron tomography to create, manipulate and reconstruct three-dimensional free electron wave packets from atomic multiphoton ionization. Examples for coherent control with bichromatic polarization-shaped pulse sequences are presented, including the generation of unusual angular momentum superposition states, carrier-envelope phase control of directional photoemission, the creation and manipulation of photoelectron vortices and electron wave packets with odd-numbered rotational symmetry as well as studies on ultrafast Rydberg and spin-orbit wave packet dynamics.

3.1 Introduction

The physical principles of quantum control by coherent light were developed more than 30 years ago [1, 2]. Essentially, the manipulation of interfering matter waves by tailored coherent light fields is the key to controlled quantum dynamics [3]. Today, coherent control by shaped femtosecond pulses is an established field in physics and chemistry that has been reviewed by many authors (see for example [4–8] and references therein). Recent developments in coherent control have been stimulated by refined experimental techniques for highly differential electron detection [9, 10] and advanced laser pulse shaping schemes [11]. For example, advances in pulse shaping has been achieved by common-path setups for full control over the vectorial electric field of ultrashort pulses [12–16], shaper-based bichromatic shaping [17] and by

---

S. Kerbstadt · K. Eickhoff · T. Bayer · M. Wollenhaupt (✉)
Institut für Physik, Carl von Ossietzky Universität Oldenburg, Carl-von-Ossietzky-Straße 9-11, 26129 Oldenburg, Germany
e-mail: Matthias.Wollenhaupt@uni-oldenburg.de

K. Yamanouchi and D. Charalambidis (eds.), *Progress in Ultrafast Intense Laser Science XV*, Topics in Applied Physics 136,
https://doi.org/10.1007/978-3-030-47098-2_3

bichromatic polarization shaping of carrier-envelope phase (CEP) stable white light supercontinua [18].

An important physical mechanism for coherent control of optically driven dynamics is based on the interference of multiple quantum pathways connecting a given initial state and a preselected final target state [7] or target wave packet. In general, the target wave packet is a coherent superposition of a set of states optically coupled to the initial state. Control over the target wave packet is achieved when the driving field connects identical states via different pathways or induces relative quantum phases within the set of interfering optically coupled states. Depending on the polarization of the driving field and the number of photons involved, different coherent control schemes can be distinguished based on the available optical control parameters and the symmetry properties of the final state [19]. Atomic and molecular multiphoton excitation or multiphoton ionization (MPI) are ideally suited to implement coherent control scenarios based on multipath interference. For example, coherent control of the momentum distributions of free electron wave packets via interfering pathways with the same number of photons (N) has been studied on atomic MPI. Single-color pulse sequences have been initially employed to manipulate the interference fringes in the photoelectron energy spectrum [20], and more recently, to create photoelectron vortices in the laser polarization plane [21].

A new twist in coherent control of atomic MPI has been introduced by the use of bichromatic polarization-tailored fields. Bichromatic fields have been used to disentangle the controlled generation of radial and angular momentum electron wave packets [22] and to localize interfering angular momentum wave packets in specific kinetic energy windows [23]. Depending on the polarization, in the single-color 'N-versus N'-photon ionization scheme depicted in Fig. 3.1a both the differential and the integral photoionization cross section can be controlled by the optical phases of the laser pulses. In contrast, the bichromatic 'N versus N' scenario shown in Fig. 3.1b permits the generation of energetically disentangled uncommon angular momentum superposition states without changing the photoionization cross section by optical phases. New possibilities for coherent control of photoionization arise when quantum paths with a different number of photons for ionization (N_1 and N_2) interfere in the continuum, as illustrated in Fig. 3.1c [23, 24]. In the scenario 'N_1 versus N_2' an odd-numbered difference $N_2 - N_1$ results in photoelectron momentum distributions (PMDs) which are, depending on the driving field polarization states, either asymmetric or have an odd-numbered rotational symmetry. As we will show below, these symmetry properties are sensitive to the CEP and therefore require a CEP-stable light source.

In this chapter, we focus on the coherent control of atomic MPI using polarization-shaped bichromatic femtosecond laser pulse sequences. In the time domain, these fields are described by

$$\mathbf{E}(t) = \mathbf{e}_{q_1}\mathcal{E}_1(t)e^{i(\varphi_1+\varphi_{ce})}e^{i\omega_1 t} + \mathbf{e}_{q_2}\mathcal{E}_2(t-\tau)e^{i(\varphi_2+\varphi_{ce})}e^{i\omega_2 t}, \tag{3.1}$$

where the envelope functions $\mathcal{E}_i(t)$ with $i = 1, 2$ characterize the shape of each individual pulse. Experimentally, the bichromatic fields are generated in the frequency

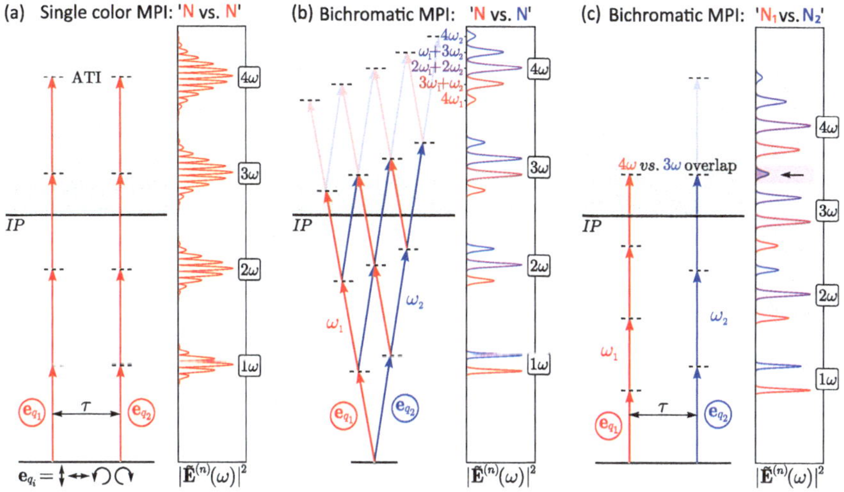

Fig. 3.1 Multiphoton ionization schemes: **a** Ionization with a time-delayed single-color double pulse sequence produces interference structures in the continuum. **b** MPI with a bichromatic field produces spectrally separated bands in the photoelectron spectrum via frequency mixing. **c** Photoelectrons from MPI with bichromatic commensurable frequency ratio ($N_1\omega$:$N_2\omega$) fields give rise to interfering angular momentum states with different parity

domain [17, 18]. The corresponding spectral representation given by the Fourier transform $\tilde{\mathbf{E}}(\omega) = \mathcal{F}\left[\mathbf{E}(t)\right](\omega)$ reads

$$\tilde{\mathbf{E}}(\omega) = \mathbf{e}_{q_1}\tilde{\mathcal{E}}_1(\omega-\omega_1)e^{i(\varphi_1+\varphi_{ce})} + \mathbf{e}_{q_2}\tilde{\mathcal{E}}_2(\omega-\omega_2)e^{i(\varphi_2+\varphi_{ce})}e^{-i(\omega-\omega_2)\tau}. \qquad (3.2)$$

Amplitude and phase modulation of the spectral bands $\tilde{\mathcal{E}}_i(\omega)$ (cf. Sect. 3.2.1) control the temporal pulse shapes $\mathcal{E}_i(t) = \mathcal{F}^{-1}[\tilde{\mathcal{E}}_i(\omega)](t)$, including the peak amplitude and duration of both pulses. The angular frequencies ω_1 and ω_2 are the center frequencies of the pulses in the sequence. These frequencies are chosen to implement either single-color $\omega_1 = \omega_2$ or bichromatic $\omega_1 \neq \omega_2$ control scenarios. The specific values of ω_i are adapted to the atomic or molecular system under investigation. To perform polarization-controlled experiments, the polarization of each pulse, described by the Jones vector $\mathbf{e}_{q_i}$, with $q_i = -1, 0, 1$, can be linear (LP) or circular (CP). The combination of the different polarizations (horizontal (HL) and vertical linear (VL)) or right handed (RCP) and left handed circular (LCP)) results in parallel linearly polarized (PLP), orthogonal linearly polarized (OLP), counter-rotating circularly polarized (CRCP) or co-rotating circularly polarized (COCP) pulse sequences. Relevant examples are shown in Fig. 3.3 in Sect. 3.2.1. Introducing a linear spectral phase $(\omega - \omega_i)\tau$ controls the temporal delay τ between the pulses. This feature allows us to study polarization-sensitive dynamics in a shaper-based pump-probe scenario. The relative phases φ_i and the CEP φ_{ce} of the field are varied to investigate phase control scenarios.

In the perturbative regime,[1] nonlinear light matter interactions involving the absorption of N photons, such as multiphoton excitation or MPI, can be studied by analyzing the higher-order nonlinear optical spectrum $\mathcal{F}\left[\mathbf{E}^N(t)\right](\omega)$ of the driving field [25]. Temporally overlapping bichromatic fields give rise to additional spectral contributions in the higher order spectrum at the frequencies $(N-k)\omega_1 + k\omega_2$ with $k = 0, \ldots, N$, as illustrated in Fig. 3.1b for $N = 2, 3$ and 4. Mathematically, these contributions result from multiple convolutions of the one-photon spectrum $\tilde{\mathbf{E}}(\omega)$ of the bichromatic field. The physical interpretation of these spectral bands is based on the absorption of $(N-k)$ photons with a frequency of ω_1 and k photons with a frequency of ω_2. The bichromatic 'N versus N' control scenario is based on the emergence of these spectral bands. Overlapping higher order spectra of different orders N_1 and N_2 lead to interferences via paths with a different number of photons, as illustrated in Fig. 3.1c. In this case, interferences in the nonlinear N_1- and N_2-order spectra, i.e. $\mathcal{F}\left[\mathbf{E}^{N_1}(t)\right] + \mathcal{F}\left[\mathbf{E}^{N_2}(t)\right]$, unlock the bichromatic 'N_1 versus N_2' control scenario which introduces the CEP as an addition control parameter.

In the following, we explore these scenarios using bichromatic polarization-shaped fields (Sect. 3.2.1) to control three-dimensional (3D) photoelectron momentum distributions (PMDs) and to study the ultrafast photoionization dynamics. Examples include coherent control by absorption of photons with different polarization and frequency (Sect. 3.3.1), directional photoemission (Sect. 3.3.2), photoelectron vortices (Sect. 3.3.3) and odd-numbered electron wave packets from bichromatic MPI (Sect. 3.3.4). The analysis of the intriguing interplay of the symmetry of the bichromatic laser field and the symmetry of the observed shaped free electron wave packet reveals that the underlying physical mechanism for control is based on the superposition of specific angular momentum states. Finally, we demonstrate the capability of the shaper-based bichromatic scheme to probe ultrafast dynamics. To this end, a linear spectral phase—resulting in a variable time delay τ—is introduced in the bichromatic pulse sequence, which allows us to map the dynamics of bound Rydberg wave packets (Sect. 3.3.5) and spin-orbit wave packets (Sect. 3.3.6) into the continuum PMD.

3.2 Experimental Techniques

In the experiments below, we employ polarization-tailored light fields for the spatial control of electron excitation/ionization dynamics and use a highly differential photoelectron detection scheme for 3D measurement of the light-induced dynamics. To this end, we combine white light polarization pulse shaping (Sect. 3.2.1) with high-resolution photoelectron tomography (Sect. 3.2.2).

---

[1]For weak interactions, i.e. when perturbation theory applies, the ground state is not depleted during the light-matter interaction. The above statement holds if there are no intermediate resonances or Stark shifts.

3.2.1 Bichromatic Polarization Shaping

Traditional $4f$ polarization pulse shaping based on a dual-layer liquid crystal spatial light modulator (LC-SLM) in the Fourier plane of a $4f$ setup [26–29] enables either independent amplitude and phase or phase and polarization pulse shaping. However, the combination of all three modes is generally not supported. To overcome this limitation, several schemes have been devised which extend conventional $4f$ pulse shapers and enable full vector control over the electric field of ultrashort laser pulses [12–15, 30, 31].

Recently, we introduced a novel shaping scheme specifically optimized to generate polarization-tailored *bichromatic* fields by amplitude modulation of an octave-spanning white light supercontinuum (WLS) [17, 18]. The experimental scheme, depicted in Fig. 3.2a, is based on a custom composite polarizer [17], which is horizontally subdivided into two parts with orthogonal transmission axes. The composite polarizer is positioned behind the LC-SLM to sculpture bichromatic amplitude profiles with orthogonal linear polarization from the input WLS via combined amplitude and phase modulation [18].

The shaper-based approach is highly versatile. All bichromatic parameters are controllable by the shaper. The bichromatic amplitude profile sketched in the inset to Fig. 3.2a [cf. (3.1)] consists of two disjoint spectral bands with individually adjustable and continuously tunable center frequencies ω_i $(i = 1, 2)$, amplitude profiles $\mathcal{A}_i(\omega)$,

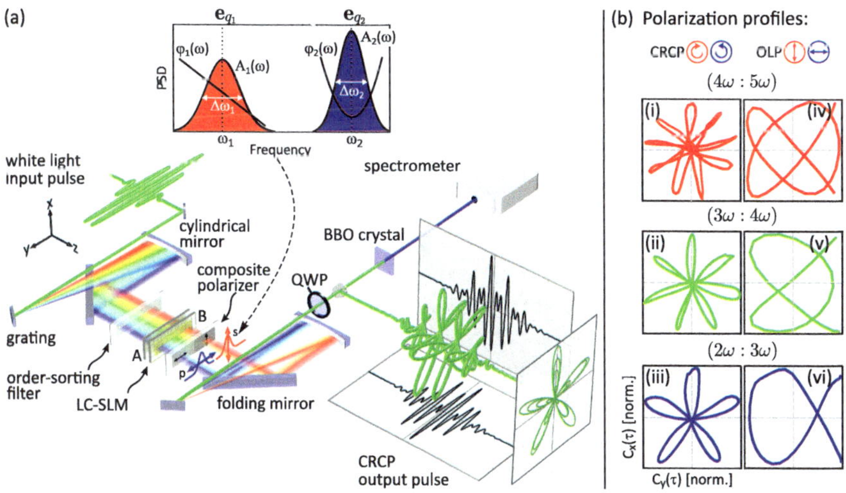

Fig. 3.2 **a** Shaping scheme based on a custom composite polarizer positioned in the Fourier plane. Fully controlled and continuously tunable bichromatic fields [see inset to (**a**)] are generated with the pulse shaper by combined bichromatic amplitude, phase and polarization shaping. **b** Measured first order cross-correlation trajectories [18] reflect the characteristic propeller-type polarization profiles of CRCP fields (i–iii) and the Lissajous-type polarization profiles of OLP fields (iv–vi) for different commensurable center frequencies

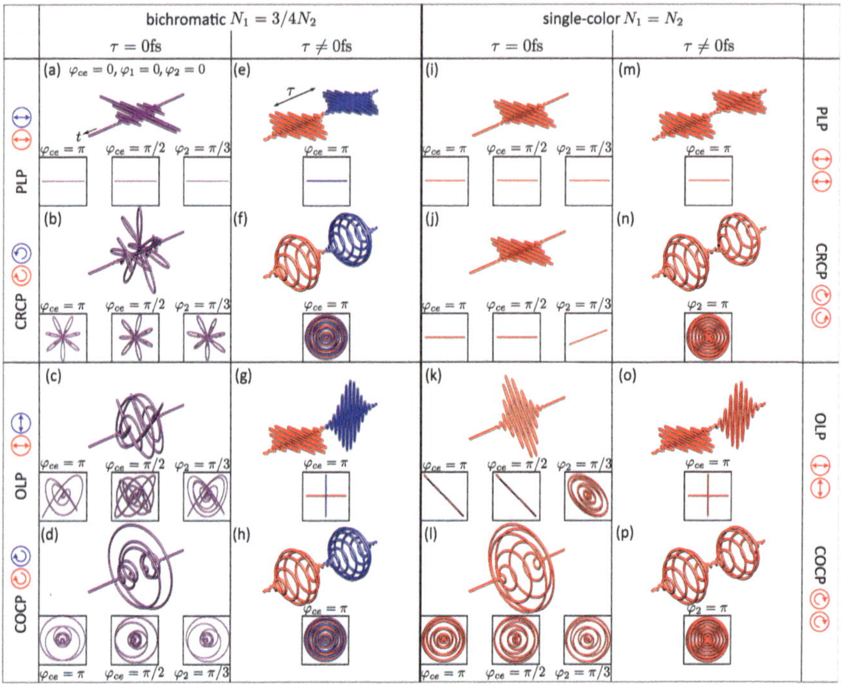

Fig. 3.3 Comparison between polarization-tailored (3ω:4ω) bichromatic (left two columns) and single-color pulse sequences (right two columns). Each class is depicted for temporal overlap ($\tau = 0$ in the first and third column) and temporal separation ($\tau \geq \Delta t$ in the second and fourth column) of the two pulses. Different rows correspond to the four polarization states PLP, CRCP, OLP and COCP. The insets show the polarization profiles of the pulses (front view) and their dependence on the relative phases φ_i and the CEP φ_{ce}

spectral widths $\Delta\omega_i$ and arbitrary phase modulation functions $\varphi_i(\omega)$. In particular, the scheme allows for precise variation of the CEP φ_{ce} and the relative phases φ_i of each color, to introduce a variable time delay τ between both colors by linear phase modulation and to select the bichromatic polarization state between PLP and OLP. Additional control of the polarization state is provided by a superachromatic quarter wave plate (QWP) at the shaper output to generate either COCP or CRCP bichromatic fields. Moreover, the scheme features built-in pulse characterization by shaper-based cross-correlation measurements. The time-delayed reference pulse is readily implemented by additional spectral amplitude and phase modulation [17, 18, 23, 32]. Finally, the shaper is used for dispersion management and in situ compression of the shaped pulses at the position of the experiment, i.e., in the interaction region of the photoelectron spectrometer. For this purpose, we employ an evolutionary algorithm to optimize the nonlinear photoelectron yield from MPI of noble gas atoms by the shaped output pulses [33, 34].

Different types of polarization-shaped bichromatic pulse sequences are illustrated in Fig. 3.3a–h and compared to their single-color counterparts in frames (i)–(p). In the case of temporally overlapping pulses ($\tau = 0$), bichromatic fields (first column) exhibit unusual polarization characteristics and vectorial field symmetries while single-color fields (third column) become linearly polarized (PLP, CRCP, OLP) or circularly polarized (COCP). For example, bichromatic ($N_1\omega$:$N_2\omega$) OLP fields with commensurable center frequencies $\omega_2 = (N_2/N_1)\omega_1$ (frame (c)) exhibit Lissajous-type polarization profiles, whereas the corresponding COCP and CRCP fields (frames (b) and (d)) feature cycloidal heart- and propeller-shaped polarization profiles, respectively [18, 35, 36]. The bichromatic polarization profiles are highly sensitive to both the relative phases φ_i and the CEP φ_{ce}, as shown in the insets. The cycloidal pulses (CRCP and COCP) are rotated by any kind of phase variation [cf. (3.3)] and also the Lissajous-pattern of the OLP pulses is strongly phase-dependent. The polarization profiles of single-color pulses are mainly affected by the relative phase which controls the spatial rotation in the CRCP case and the ellipticity in the OLP case. Only in the COCP case, the pulse is rotated by both the relative phase and the CEP. These differences between bichromatic and single-color fields has profound consequences for light-matter interactions and the optical parameters available for control, as will be discussed in Sects. 3.3.2 and 3.3.4. In the case of temporally separated pulses, the respective polarization states are meaningfully defined for bichromatic and also for single-color fields. Such pulse sequences are used e.g. in Sect. 3.3.3 for the creation of single-color electron vortices and further in Sect. 3.3.5 and Sect. 3.3.6 for bichromatic pump-probe studies on Rydberg and spin-orbit wave packets, respectively.

Traditionally, commensurable bichromatic fields are generated by superposition of harmonic beams using an interferometer [37, 38]. These fields are inherently CEP-insensitive, leaving the relative phase between the colors as the central parameter for coherent control. In contrast, shaper-generated bichromatic fields depend on both the relative phases φ_i and the CEP φ_{ce}, each with different sensitivities. For example, in the case of cycloidal ($N_1\omega$:$N_2\omega$) fields, any phase variation translates into a spatial rotation of the laser field about its propagation axis by the angles

$$\begin{aligned} \alpha^{cr} &= \frac{N_2 - N_1}{N_2 + N_1}\varphi_{ce} + \frac{N_2}{N_2 + N_1}\varphi_1 - \frac{N_1}{N_2 + N_1}\varphi_2 \\ \alpha^{co} &= \varphi_{ce} + \frac{N_2}{N_2 - N_1}\varphi_1 - \frac{N_1}{N_2 - N_1}\varphi_2, \end{aligned} \tag{3.3}$$

where α^{cr} and α^{co} denote the rotation angles of CRCP and COCP fields, respectively [18]. On the one hand this implies that all optical phases are utilizable as control parameters in the shaper-based scheme. On the other hand, stabilization of all phases – in particular the CEP – is crucial if shaper-generated bichromatic fields are employed for the spatial control of quantum dynamics. Otherwise phase fluctuations, leading to random spatial rotations of the cycloidal fields, will average out the phase-sensitive interferences.

3.2.2 Photoelectron Tomography

Coherent control of ultrafast dynamics using polarization-shaped pulses exploits the vectorial character of light-matter interactions. This opens the door to the manipulation of spatial aspects of quantum phenomena such as asymmetries in the wave function of a quantum system or the directionality of quantum processes. In order to utilize the full potential of spatial control, the application of 3D detection techniques providing vectorial information on the dynamics is crucial as well. Among the various detection schemes, based on the measurement of either photons or massive particles, the measurement of photoelectrons is particularly attractive due to its high detection efficiencies and the wealth of information provided by the 3D PMD. Different methods have been devised to measure the PMD, yielding highly differential information on the processes under investigation. For example, fundamental aspects of light-matter interactions have been studied using cold target recoil ion momentum spectroscopy (COLTRIMS) [10]. The COLTRIMS technique permits the reconstruction the 3D PMD by coincidence measurement of photoelectrons and photoions with angular resolution. Another well-established method to measure the photoelectron energy and angular distribution is velocity map imaging (VMI) [9]. The basic concept is depicted in Fig. 3.4a. In VMI, the PMD is imaged by electrostatic lenses onto a position-sensitive 2D detector consisting of a multi-channel plate (MCP) and a phosphor screen. The image of the screen is recorded by a charge-coupled device (CCD) camera. If the PMD is cylindrically symmetric, a single 2D projection is

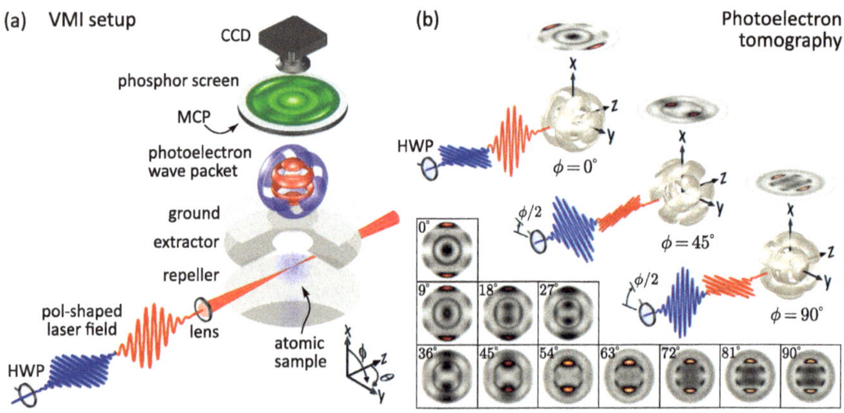

Fig. 3.4 Experimental scheme for tomographic imaging of free electron wave packets. **a** shows a VMI setup consisting of an assembly of electrodes (repeller, extractor, ground) to image the photoelectron wave packet, generated by the interaction of the sample with the laser pulse, onto a position-sensitive MCP detector. The 2D images of the detector are recorded by a CCD camera. **b** illustrates the tomographic procedure applied to reconstruct the 3D photoelectron momentum distribution. By rotation of the laser pulse about its propagation axis using a half wave plate, different 2D projections of the electron density under various angles ϕ are measured. From the series of 2D images, the 3D density is retrieved using a tomographic algorithm

sufficient to retrieve the full 3D information via Abel inversion [39]. However, in general PMDs created by polarization-shaped pulses exhibit no such symmetry. In this case, tomographic techniques, adapted from medical applications, are employed to reconstruct the 3D PMD [40, 41]. The working principle of photoelectron tomography is illustrated in Fig. 3.4b. Numerous 2D projections of the PMD are measured under different projection angles by rotating the ionizing laser pulse about its propagation axis using a superachromatic half wave plate (HWP). From the acquired set of 2D images, the 3D PMD is reconstructed using tomographic procedures such as the Fourier slice algorithm [40, 42, 43] or the backprojection algorithm [41, 42, 44]. Recent applications of the tomographic method comprise the imaging of molecular orbitals [43], the creation and measurement of designer free electron wave packets [45], the discrimination of chiral molecules via the multi-photon photoelectron circular dichroism [46–48], the characterization of elliptically polarized high-harmonic radiation and the time-resolved investigation of laser-matter interaction inside transparent materials [44]. Further applications are reported in this chapter, including the generation and detection of unusual angular momentum superposition states [22] (Sect. 3.3.1), free electron wave packets with odd rotational symmetries [49] (Sect. 3.3.4), photoelectron vortices [21, 50] (Sect. 3.3.3) and the spatial imaging of spin-orbit wave packet dynamics [51] (Sect. 3.3.6).

3.3 Control of Free Electron Wave Packets

In this section, we present various applications of the experimental techniques introduced above. First we discuss different control scenarios to sculpture free electron wave packets from MPI of atomic systems employing the shaper-generated polarization-tailored bichromatic fields. Subsequently, we describe two time-resolved studies in which the bichromatic optical scheme was utilized for the observation of atomic Rydberg and spin-orbit wave packet dynamics, respectively.

3.3.1 Control by Frequency Mixing

In the first control scenario, we use bichromatic fields with incommensurable center frequencies ω_1 and ω_2 to control the 3D PMD from atomic MPI. Specifically we employ temporally overlapping OLP pulses as prototypes for polarization-shaped bichromatic fields. The OLP pulses consist of an s-polarized red band centered around $\omega_1 = 2.28$ rad/fs (826 nm) and a p-polarized blue band centered around $\omega_2 = 2.45$ rad/fs (769 nm). Both colors have the same temporal shape $\mathcal{E}_1(t) = \mathcal{E}_2(t) \equiv \mathcal{E}(t)$ with a spectral width of $\Delta\omega = 0.035$ rad/fs which corresponds to a pulse duration of $\Delta t = 80$ fs. By setting $\tau = 0$ and decomposing each linearly polarized field into a superposition of two counter-rotating circularly polarized components, the laser field in (3.1) is rewritten as

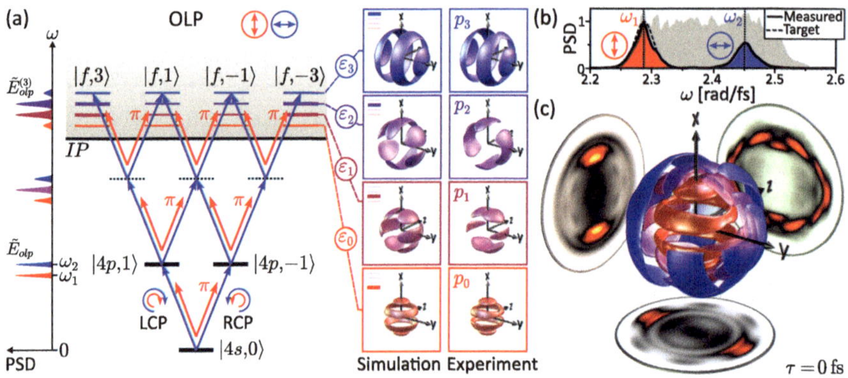

Fig. 3.5 Multipath control of MPI by frequency mixing between two spectral bands of different color and ellipticity. **a** Excitation scheme for (1+2) REMPI of potassium (K) atoms using temporally overlapping bichromatic OLP pulses with incommensurable center frequencies. Both colors are decomposed into an LCP and an RCP component. Multiple ionization pathways from intra-pulse frequency mixing are disentangled by energetic separation of the target states $|f, m_j\rangle$ in the continuum. This gives rise to unusual superpositions of angular momentum eigenstates, as illustrated in the insets (left columns). **b** Measured bichromatic amplitude profile used in the experiment. **c** Tomographically reconstructed photoelectron density. Decomposition of the PMD into different momentum shells p_n, corresponding to the four ionization channels ε_n, yields the individual photoelectron wave packets shown in the insets (right column)

$$\mathbf{E}_{olp}(t) = \mathbf{e}_{-1}\,\mathcal{E}(t)[e^{i(\omega_1 t+\pi)} + e^{i\omega_2 t}] + \mathbf{e}_1\,\mathcal{E}(t)[e^{i\omega_1 t} + e^{i\omega_2 t}]. \tag{3.4}$$

Here, all phases (φ_1, φ_2 and φ_{ce}) were set to zero. The extra phase of π in the RCP component of the red pulse arises due to the 90° spatial rotation of the red pulse with respect to the blue. Unlike single-color OLP pulses, which create linear polarization when coinciding in time (see Fig. 3.3k), the bichromatic field in (3.4) describes a complex, Lissajous-type polarization profile (cf. Fig. 3.3c). In a multiphoton excitation scenario, temporal overlap of both colors enables frequency mixing between photons of different color and opposite circularity. All possible photon combinations leading to the continuum create a manifold of ionization pathways which, in a single-color scenario, interfere at the same photoelectron energy. In a bichromatic frequency mixing scenario, however, multiple pathways are disentangled by energetic separation of different angular momentum target states. This provides the basis to synthesize free electron wave packets by selective superposition of angular momentum eigenstates.

To demonstrate bichromatic control of the 3D PMD via incommensurable frequency mixing, we investigate (1+2) resonance-enhanced MPI (REMPI) of potassium (K) atoms as a model system. The corresponding excitation scheme is depicted in Fig. 3.5a. The center frequency ω_2 of the blue band is tuned to the K $4s \rightarrow 4p$ resonance while the red band centered at ω_1 is completely non-resonant. According to the dipole selection rules[2] $\Delta\ell = 1$ and $\Delta m = \pm 1$ for $\sigma^{\pm}$ transitions, the bichromatic OLP pulse gives rise to numerous interfering ionization pathways. Three-photon ionization connects the ground state $|4s, 0\rangle$ to different target states $|f, m_j\rangle$, with $m_j = 3, 1, -1$ and -3, in the continuum. In the perturbative limit, the kinetic energy distribution of the created photoelectron wave packets is essentially determined by the third-order spectrum of the ionizing laser pulse [22, 25]. The third-order spectrum $\tilde{E}^{(3)}_{olp}(\omega) = \mathcal{F}[E^3_{olp}(t)](\omega)$ of the OLP field $E_{olp}(t) = |\mathbf{E}_{olp}(t)|$ in (3.4) exhibits four contributions centered around $\omega'_n = (3 - n)\,\omega_1 + n\,\omega_2$, with $n = 0, \ldots, 3$. As a result, we obtain four energetically separated ionization channels centered around the photoelectron kinetic energies $\varepsilon_n = \hbar\omega'_n - IP$, where IP denotes the atomic ionization potential. The two outermost channels at ε_0 and ε_3 correspond to single-color three-photon ionization by the red and the blue pulse, respectively. The inner channels at ε_1 and ε_2 correspond to intra-pulse frequency mixing of both colors. In [22] it was shown, that the general photoelectron wave function of the n-th ionization channel can be expressed as

$$\psi_n(\varepsilon, \theta, \phi) \propto \gamma(\varepsilon - \varepsilon_n) \sum_{j=0}^{3} a_{nj} Y_{3,m_j}(\theta, \phi), \tag{3.5}$$

which describes a superposition of the angular momentum target states $|f, m_j\rangle$. The corresponding angular distributions are given by the spherical harmonics $Y_{\ell,m}(\theta, \phi)$. The shape function $\gamma(\varepsilon)$ determines the photoelectron kinetic energy distribution in dependence of the bichromatic amplitude profile. In the case of two identically shaped colors, each channel has the same distribution centered around the corresponding energy ε_n and given by the third-order spectrum of the envelope function $\tilde{\mathcal{E}}(\omega) = \mathcal{F}[\mathcal{E}(t)](\omega)$, with $\omega = \varepsilon/\hbar$. The amplitudes a_{nj} are determined by the interference of all ionization pathways leading to the respective target state $|f, m_j\rangle$ at energy ε_n. These amplitudes are given by the coherent sum over the pathway weights which are products of the dipole couplings along each pathway and the frequency mixing amplitude of the associated third order optical spectrum. Thus, the final target state population results from an interplay of dipole selection rules, addressed via the polarization state of the bichromatic field, and intra-pulse frequency mixing.

Figure 3.5b shows the measured bichromatic amplitude profile of the OLP field used in the experiment. The amplitude ratio was set to $A_1/A_2 \approx 2$ to promote the non-resonant pathways starting with a red photon. The tomographically reconstructed 3D PMD created by REMPI of K atoms using this field is shown in Fig. 3.5c in momentum (p) representation. Four nested photoelectron wave packets are observed in different radial shells centered around the momenta p_n, corresponding to the four ionization channels (ε_n). For better visibility, all four contributions are displayed separately in the central insets (right column) where the experimental results are compared to numerical calculations based on (3.5) (left column). The inner- (p_0) and

[2]Transitions with $\Delta\ell = -1$ are possible as well. However, for simplicity we focus on the $\Delta\ell = 1$ case which, in addition, is favored by propensity rules [52].

outermost (p_3) contributions from single-color ionization exhibit rotated $|f, m = 0\rangle$-symmetry. In accordance with the associated laser polarizations, the wave packet in the p_0-channel is aligned vertically along the x-axis while the wave packet in the p_3-channel is aligned horizontally along the y-axis. The wave packets in the two inner channels (p_1 and p_2), created by intra-pulse frequency mixing, are also aligned perpendicular to each other – best discernible in the simulations. These rather unusual wave packets are mainly composed of the two 'counter-rotating' states $|f, \pm 3\rangle$ with only minor contributions of the states $|f, \pm 1\rangle$ [22]. The resulting superposition creates an almost pure standing wave pattern with quasi c_6 rotational symmetry in azimuthal (ϕ) direction (cf. Sect. 3.3.4).

In conclusion, the results presented in this section exemplify the power of polarization-shaped bichromatic fields with incommensurable center frequencies to sculpture the 3D PMD from atomic MPI by creation of unusual angular momentum superposition states. In the prototypical OLP case discussed here, we demonstrated the design of a rotated 'f within f' wave packet and a pair of wave packets with quasi c_6 rotational symmetry. Both examples represent free electron wave packets which can not be created by single-color MPI. In a second experiment based on bichromatic CRCP pulses, we demonstrated the selective creation of individual angular momentum eigenstates [22]. It should be noted, that the interference of frequency mixing pathways of the same order is insensitive to the CEP φ_{ce}. However, control of the 3D PMD via the relative optical phase $\Delta\varphi = \varphi_2 - \varphi_1$ between the two colors is possible if the different ionization channels overlap energetically, i.e., if the bandwidth of the N-th order frequency mixing contribution is larger than the channel separation. In general, the underlying physical mechanism is based on the interplay of ionization pathway selection via quantum mechanical selection rules for optical transitions and intra-pulse frequency mixing of spectral bands with different ellipticity.

3.3.2 Control of Directional Photoemission

Coherent control of atomic and molecular MPI in the perturbative regime is generally based on the interference of different multiphoton excitation pathways. Different control scenarios may be distinguished by the parity of the involved target states. If the interfering target states are of the same parity, control of both the integral (yield) and differential (angular distribution) photoionization cross section [19] is possible, but the final-state wave function is always spatially (left-right) symmetric. An example of this type of control is presented in Sect. 3.3.3, where we discuss the creation of even-numbered photoelectron vortices by MPI with single-color CRCP pulse sequence [21, 50]. In contrast, spatial asymmetries in the photoelectron angular distribution are induced by the interference of continuum states with opposite parity. These states are addressed by MPI with different numbers of photons [19, 37, 38, 53], e.g., via N_1- versus N_2-photon ionization with $N_2 = N_1 + 1$. Recent examples for this kind of spatial control exerted on the final state wave function comprise the phase-sensitive directional emission of photoelectrons along the laser polarization in CEP-stable few-cycle pulses [37, 38, 54, 55], bichromatic (ω:2ω) fields [38, 53, 56] and bichromatic (7ω:8ω) PLP fields [23].

In few-cycle MPI, CEP-sensitive N_1- versus N_2-photon interferences arise naturally in the energetic overlap region between the threshold and adjacent above threshold ionization (ATI) channels. In this case, a multitude of quantum pathways lead to the same photoelectron energy. Their intricate interference conceals the underlying physical mechanisms for control and makes the precise manipulation of the final state wave function by forward design of the fundamental field difficult [23, 55, 57]. Recently, we showed that shaper-generated bichromatic fields with center frequencies ω_i tailored to the order of the MPI process allow to create almost pure N_1- versus N_2-photon interferences in a selected energy window [23]. In this experiment, we employed (7ω:8ω) PLP fields to control asymmetries in the photoemission from high-order MPI of xenon (Xe) atoms. It was demonstrated that the shaper-based bichromatic approach preserves the CEP-sensitivity of the excitation and provides refined optical control over the multipath interference of the target states. In particular, the efficiency of the involved ionization pathways was individually adjustable via the bichromatic amplitude ratio and the quantum phases of the interfering target states were controlled via the spectral phases of the two colors.

Here, we present experimental results on the control of spatial asymmetries in the PMD from MPI of sodium (Na) and K atoms using shaper-generated (3ω:4ω) PLP fields. The corresponding excitation scheme is displayed in Fig. 3.1c. Commensurable center wavelengths $\lambda_1 = 880$ nm (red pulse) and $\lambda_2 = \frac{3}{4}\lambda_1 = 660$ nm (blue pulse) are chosen to induce interference of photoelectron wave packets from 3- and 4-photon ionization of the alkali atoms. The former gives rise to f ($m = 0$)-type wave packets (odd parity), whereas the latter creates g ($m = 0$)-type wave packets (even parity), as illustrated in the top insets to Fig. 3.6a, b. The total photoelectron wave function in the relevant energy window can be written as [24]

$$\psi_{dir} \propto \psi_{3,0} + i\,\psi_{4,0}\,e^{-i\Delta\varphi}, \tag{3.6}$$

with $\Delta\varphi = 4\varphi_1 - 3\varphi_2 + \varphi_{ce} + \tau\omega$. As shown in the top insets to Fig. 3.6c, d, the final-state wave function is asymmetric along the y-axis i.e. in the polarization direction of the PLP fields. The asymmetry is determined by the relative quantum phase $\Delta\varphi$ between the interfering photoelectron wave packets. In a first proof-of-principle experiment performed on K atoms, we utilize this asymmetry as a sensitive probe to verify the coherence properties of the shaper-generated bichromatic (3ω:4ω) fields by investigating the temporal, spatial and energetic overlap of the created photoelectron wave packets. To this end, the bichromatic fields are focused into the VMI spectrometer (focus intensity $I = 2 \times 10^{12}$ W/cm^2) and 2D photoelectron images are measured for different values of the CEP φ_{ce}, while setting $\varphi_i = 0$ and $\tau = 0$. In order to analyze the photoelectron asymmetry, 2D sections through the PMD in the polarization plane (x-y-plane) are retrieved via Abel inversion using the pBASEX algorithm [39]. All 2D sections are energy-calibrated following the procedure described in [58]. Figure 3.6a, b display the g- and f-type single-color photoelectron spectra, obtained by MPI with only the red and the blue pulse, respectively in

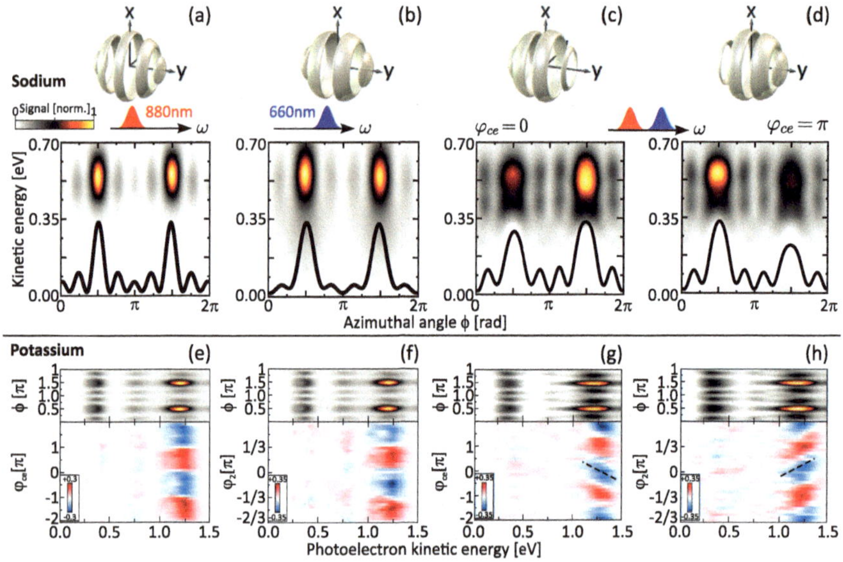

Fig. 3.6 Phase-sensitive directional photoemission by MPI of alkali (potassium (K) and sodium (Na)) atoms with (3ω:4ω) PLP fields. **a** The center frequency of the red band is chosen such that a 4-photon ionization into the g-type continuum (even parity) is induced and **b** the blue band triggers a 3-photon ionization into the f-type continuum. **c, d** The asymmetry of the coherent superposition of the continuum states with opposite parity is inverted by switching the CEP from $\varphi_{ce} = 0$ to $\varphi_{ce} = \pi$. The measured data in (**a**)–(**d**) are based on MPI of Na atoms. Calculated 3D wave packets are plotted for comparison. To evaluate the influence of the relative phase φ_2 and the CEP φ_{ce}, energy-resolved 2D asymmetry maps are generated [23] from MPI of K atoms for $\tau = 0$ fs (**e**)–(**f**) and for $\tau = -25$ fs (**g**)–(**h**)

polar representation. The spectra reveal that the created photoelectron wave packets are localized in a kinetic energy window centered around $\varepsilon \approx 0.5$ eV. The spectrum from bichromatic MPI with $\varphi_{ce} = 0$ is shown in Fig. 3.6c. Due to the interference of the two wave packets with opposite parity, the photoemission is localized around $\phi = 3\pi/2$, i.e., in negative y-direction. By switching the CEP to $\varphi_{ce} = \pi$ this asymmetry is inverted, as shown in Fig. 3.6d. Here the photoemission maximum is observed around $\phi = \pi/2$.

According to (3.6), the spatial asymmetry of the final-state wave function is sensitive to all optical phases, however, with different sensitivities. For example, a CEP variation of $\Delta\varphi_{ce} = \pi$ and a variation of the relative phase φ_2 by $\Delta\varphi_2 = \pi/3$ result in an equivalent inversion of the asymmetry. To demonstrate this equivalence, we performed an additional experiment on K atoms in which we studied the photoelectron asymmetry under continuous variation of the CEP and the relative phase of the blue pulse. The results are presented in Fig. 3.6e, f. The bottom frames show energy-resolved asymmetry maps obtained by anti-symmetrization and angle-integration of the retrieved 2D sections [23, 57]. The top frames show phase-averaged photoelectron spectra for the assignment of signal contributions in the maps. In the case of K atoms, the overlap of photoelectron wave packets from 3- and 4-photon ionization is observed in a kinetic energy window centered around $\varepsilon \approx 1.3\,\text{eV}$. In this window, we observe alternating patterns of destructive (negative asymmetry) and constructive (positive asymmetry) interference upon variation of the respective optical phase. Scaling the φ_2-axis in panel (f) by a factor of 1/3 and comparing the result to the φ_{ce}-map in panel (e) clearly shows the equivalence between the effect of the two phases on the photoelectron asymmetry.

Equation (3.6) furthermore predicts that the photoelectron interference is sensitive to the time delay τ between the two colors. Experimentally, this time delay is introduced by linear spectral phase modulation. While the asymmetry observed at $\tau = 0\,\text{fs}$ (see Fig. 3.6e, f) is energy-independent (flat) in the relevant energy window, for $\tau = -25\,\text{fs}$ the interference pattern acquires an energy-dependent linear tilt with negative slope in the φ_{ce}-asymmetry map shown in Fig. 3.6g. The physical mechanism behind this tilt was discussed in [23, 55, 57] in terms of the additional chirp-dependent quantum phase. In the bichromatic MPI scenario considered here, the slope of the tilt is inverted when the relative phase φ_2 is varied instead of the CEP (cf. Fig. 3.6h). This result verifies the negative sign of the φ_2-contribution to $\Delta\varphi$ in (3.6) relative to the CEP-dependent term. As a consequence, cycloidal bichromatic fields (cf. Sect. 3.2.1) rotate in opposite directions by variation of φ_{ce} and φ_2, respectively, as is described by (3.3). The interplay between the phase-sensitivity of cycloidal bichromatic fields and the created photoelectron wave packets is discussed in more detail in Sect. 3.3.4.

3.3.3 Single Color Electron Vortices

Ionization with two time-delayed CRCP pulses creates free electron wave packets, that are shaped like an Archimedean spiral in the laser polarization plane [21, 59]. Inspired by the helical interference structures, the resulting PMD was termed an 'electron vortex' by Starace and coworkers [59]. This notion of an electron vortex has to be distinguished from the traditional definition of vortex states in quantum systems [60, 61], derived from the hydrodynamic formulation of quantum mechanics [62]. Vortex states, which are defined by their non-vanishing azimuthal probability current density of the wave function

$$\mathbf{j}(\mathbf{r}) \propto \Im\left[\psi^*(\mathbf{r})\nabla\psi(\mathbf{r})\right] = |\psi(\mathbf{r})|^2\nabla \arg\psi(\mathbf{r}), \tag{3.7}$$

are subject of intense studies in collision physics [61, 63] and the generation of electron vortex beams [64, 65]. For general N_1- versus N_2-photon ionization, using CRCP laser pulses with a time delay τ the probability current density of the resulting photoelectron wave packet reads

$$\mathbf{j}(\mathbf{r}) \propto \frac{1}{2i}|\psi|^2\left(\frac{i\tau}{\hbar}\mathbf{e}_\varepsilon + \frac{i(N_1 - N_2)}{\varepsilon\sin(\theta)}\mathbf{e}_\phi\right), \tag{3.8}$$

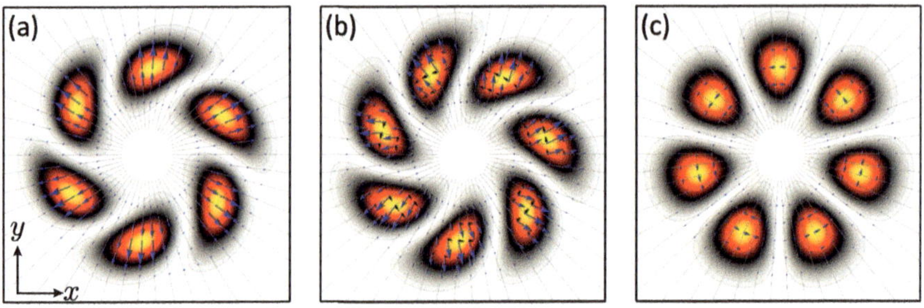

Fig. 3.7 Contour plots of simulated photoelectron wave packets along with their lateral probability current $\mathbf{j}(x, y)$ indicated by blue arrows. **a** The spiral-shaped photoelectron wave packet from CRCP single-color photoionization with a pulse sequence ($\tau = 7$ fs) has c_6 rotational symmetry and a purely radial probability current, collinear to the gray radial auxiliary lines. **b** The electron vortex form bichromatic (3ω:4ω) ionization with c_7 rotational symmetry has both radial and azimuthal contributions. The azimuthal components are indicated with smaller black arrows. **c** The probability current of the bichromatic c_7 free electron wave packet from (3ω:4ω) cycloidal photoionization ($\tau = 0$) is purely azimuthal

where the azimuthal component vanishes for the single color case, i.e., $N_1 = N_2$ (for notation see [24]). According to (3.7) this type of vortex manifests in the properties of the quantum mechanical phase arg $\psi(\mathbf{r})$. In contrast, free electron vortices as discussed in [21, 59, 66, 67] are spiral-shaped angular distributions of photoelectron wave packets characterized by the shape of their electron density. The creation of even-armed photoelectron vortices by single-color MPI of K atoms was demonstrated experimentally in [21, 50]. The probability current $\mathbf{j}(\mathbf{r})$ of these objects has no azimuthal component (see (3.8) and arrows in Fig. 3.7), i.e., zero vorticity in the hydrodynamic sense. However, odd-numbered photoelectron wave packets from bichromatic MPI exhibit an azimuthal current, both for $\tau \neq 0$ fs (Fig. 3.7b) and for $\tau = 0$ fs (Fig. 3.7c). Experimental results on these two scenarios are presented in this section.

We start with the discussion of vortex-shaped electron wave packets with an even number of vortex arms, created by REMPI with time-delayed single-color CRCP pulse sequences. In general, the creation of photoelectron vortices relies on the superposition of two time-delayed free electron wave packets with different magnetic quantum numbers m. During its time evolution, the first wave packet accumulates an energy-dependent quantum mechanical phase of $-\frac{\varepsilon}{\hbar}\tau$. The interference of both free electron wave packets results in Ramsey-type fringes in the energy distribution [20], i.e. in the radial direction in the PMD. In the first experiment, we use WLS input pulses provided by an argon (Ar) filled hollow-core fiber. The WLS has a spectral width of $\Delta\omega \simeq 0.40$ rad/fs and a pulse duration of about $\Delta t \simeq 7$ fs (see inset Fig. 3.8a, b). Shaper-generated time-delayed CRCP WLS pulse sequences were employed to investigate (1+2) REMPI of K atoms, depicted in Fig. 3.1a. According to the selection rule $\Delta m = \pm 1$ for CP pulses, each pulse of the sequence ionizes the atom via a single ionization pathway. Three-photon ionization with an LCP pulse

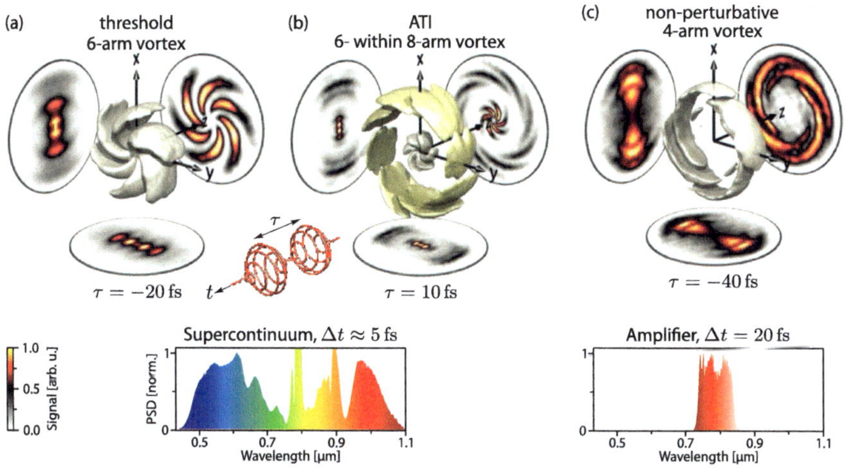

Fig. 3.8 Tomographically reconstructed electron wave packets (upper panel) and laser pulse spectra used in the measurements (bottom panel). **a** 6-armed electron vortex with clockwise sense of rotation from perturbative (1+2) REMPI using a CRCP sequence of WLS pulses. **b** The absorption of an additional photon in the continuum (ATI) gives rise to the 8-armed vortex. **c** Photoelectron vortex with c_4 rotational symmetry from non-perturbative photoionization with a sequence of two CRCP π-pulses

preceding an RCP pulse creates a coherent superposition of f-type photoelectron wave packets with quantum numbers $m = \pm 3$:

$$\psi_{c_6} \propto \psi_{3,-3} + e^{-i\frac{\varepsilon}{\hbar}\tau}\psi_{3,3}. \tag{3.9}$$

In the angular direction, the interference of the two wave functions $\psi_{3,\pm 3}$ causes a c_6 rotational symmetry of the wave packet in the polarization plane [21]. The reconstructed and energy-calibrated 3D PMD is shown in Fig. 3.8a for a time delay of $\tau = -20$ fs. The maxima of the PMD in the polarization plane form an Archimedean spiral which rotates in clockwise direction for increasing values of ε (radial direction). Taking into account the handedness conventions and the orientation of the plotted sections, our experimental findings are consistent with the theoretical prediction in [59]. Slight deviations from a perfect c_6 symmetry, expressed by the inhomogeneous intensity distribution among the spiral arms, are explained by minor contributions of the $4p$-state, perturbatively excited by the first pulse and ionized by the second pulse [50]. In the next step, we investigated free electron vortices in the ATI spectrum. The experimental result for a time delay of $\tau = 10$ fs is shown in Fig. 3.8b. Analogous to the c_6 electron vortex created by three photon absorption described by (3.9), the final-state wave packet in the first ATI ($m = \pm 4$) is an electron vortex with c_8 rotational symmetry due to the additionally absorbed photon [50]. Compared to the photoelectron vortex presented in Fig. 3.8a, the sense of rotation is reversed by the inverted sign of τ which reverses the pulse ordering. In the second experiment, we investigated electron vortices created by non-perturbative (1+2) REMPI of K atoms using CRCP pulse sequences from the amplifier. The spectral width of the pulses is $\Delta\omega = 0.14$ rad/fs (cf. inset Fig. 3.8c) corresponding to a pulse duration of $\Delta t \simeq$ 20 fs. While perturbative excitation leaves the ground state population essentially unchanged, the population dynamics during non-perturbative interaction of the first pulse with the quantum system changes the initial conditions for the second laser pulse. Non-perturbative excitation with an RCP π-pulse depletes the ground state and stores the population exclusively in the excited $4p$ $(m = -1)$ state. This excited state is ionized by absorption of two photons from the subsequent LCP pulse, leading to a wave packet

$$\psi_{c_4} \propto \psi_{3,1} + e^{-i\frac{\varepsilon}{\hbar}\tau}\psi_{3,-3}. \tag{3.10}$$

with c_4 rotational symmetry in the polarization plane. Our results show that photoelectron vortices are sensitive to the ionization dynamics, in particular to the population of intermediate resonances. Using one of the free electron wave packets as a reference, the evaluation of the spectral phase of the interfering electron wave packets yields the relative quantum mechanical phase of the other free electron wave packet. This type of quantum state holography provided a direct measurement of the wave function (including the relative phase) rather than its probability density [50].

3.3.4 Odd-Numbered Electron Wave Packets from Bichromatic MPI

The full potential of coherent control with bichromatic fields unfolds when ultrashort phase-stable $(N_1\omega{:}N_2\omega)$ COCP and CRCP laser pulses are used for photoionization. Such fields exhibit cycloidal polarization profiles [24, 35, 36], with a field symmetry of $\mathcal{S}_{opt} = (N_1 \mp N_2)/\,\mathrm{gcd}\,(N_1, N_2)$ determined by the center frequency ratio $\omega_2/\omega_1 = N_1/N_2$ [18]. The plus and minus sign correspond to CRCP and COCP fields, respectively, and gcd denotes the greatest common divisor. If the fields of both colors overlap in time, their individual polarization characteristics are imprinted in the cycloidal polarization profile of the resulting pulse. For example, temporally overlapping CRCP bichromatic fields $(\tau = 0)$ create propeller-type pulses, as visualized in Fig. 3.3b on the example of a 7-leafed $(3\omega{:}4\omega)$ CRCP pulse. In comparison, temporally overlapping single-color CRCP pulse sequences are linearly polarized, as also shown in Fig. 3.3j and discussed in Sect. 3.3.3. By employing such cycloidal fields for photoionization, 3D control over spatial symmetries of the free electron wave packets is achieved. The underlying physical mechanism is based on the interference of electronic wave functions with different angular momenta addressed by ionization pathways with a different number of photons. In recent experiments $(\omega{:}2\omega)$ CRCP pulses with a 3-fold propeller-type polarization profile were employed for the creation of photoelectron wave packets with 3-fold rotational symmetry by strong-field ionization of Ar atoms [68–70]. In further theoretical studies, the creation of even-

and odd-numbered photoelectron vortices by ionization of atoms [71] and molecular ions [66, 72] with circularly polarized bichromatic laser fields [18] were investigated.

In the following, we present an experimental scheme to create and manipulate 3D free electron wave packets with arbitrary rotational symmetry by combining bichromatic white light shaping [18] with high-resolution photoelectron tomography [40] as shown in Fig. 3.4. We exert full spatial control over photoelectron wave packets by designing specific energy and angular momentum superposition states. For this purpose, we use shaper-generated CEP-stable bichromatic (3ω:4ω) COCP and CRCP femtosecond laser pulses to create 7-fold rotationally symmetric and asymmetric PMDs by MPI of Na atoms (the pulse parameters are given in Sect. 3.3.2). In the CRCP case, with a red LCP pulse and a blue RCP pulse, the final-state wave function is the superposition of two counter-rotating torus-shaped wave packets:

$$\psi_{cr} \propto \psi_{3,-3} + i\,\psi_{4,4}\,e^{-i\Delta\varphi}, \tag{3.11}$$

with $\Delta\varphi = 4\varphi_1 - 3\varphi_2 + \varphi_{ce}$. The superposition state in (3.11) describes a spherical standing wave with $\mathcal{S}_{wp} = N_2 + N_1 = 7$ lobes in the laser polarization plane (x-y-plane), i.e., with c_7 rotational symmetry, as shown in Fig. 3.9a. Examples for other frequency ratios are depicted in Fig. 3.10. As described by (3.2) and (3.3), the shaper-generated cycloidal fields depend on both the CEP and the relative phases between the fields of the two colors. The optical phases enter with different weights into the relative quantum phase $\Delta\varphi$ between the interfering wave functions. Switching from CRCP to COCP allows us to localize the photoelectrons in one half of the polarization plane. The resulting photoelectron wave packet is described by the wave function

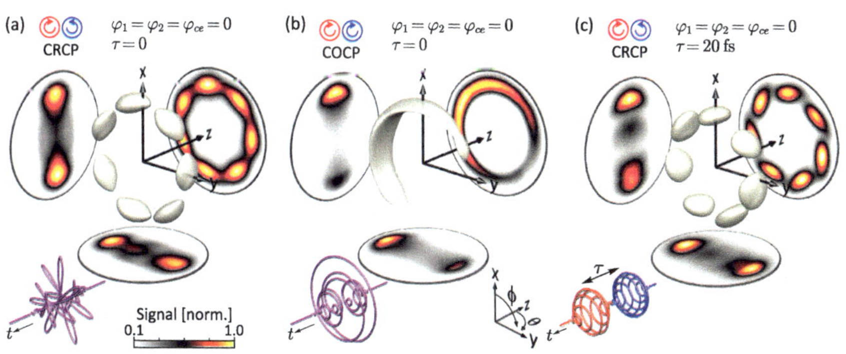

Fig. 3.9 Odd-numbered photoelectron wave packets from MPI of Na atoms with (3ω:4ω) CRCP and COCP fields, reconstructed using high-resolution photoelectron tomography. **a** Photoelectron wave packet with CEP-sensitive c_7 rotational symmetry from MPI of Na atoms with a (3ω:4ω) CRCP field. **b** Switching to a COCP field allows to localize the photoelectron in one half of the polarization plane. The photoelectron wave packet is crescent-shaped and asymmetric. **c** Introducing a time delay between both colors of the (3ω:4ω) CRCP field creates a 7-armed photoelectron vortex, retaining the c_7 symmetry despite the circular symmetry of the ionizing laser field

$$\psi_{co} \propto \psi_{3,3} + i\,\psi_{4,4}\,e^{-i\Delta\varphi}, \tag{3.12}$$

with $\mathcal{S}_{wp} = N_2 - N_1 = 1$. The resulting crescent-shaped electron density is depicted in Fig. 3.9b. Introducing a time delay between the two colors results in a linear tilt of the lobes of the standing wave. The corresponding wave function

$$\psi_{vor} \propto \psi_{3,-3} + i\,\psi_{4,4}\,e^{-i\Delta\varphi}e^{i\varepsilon\tau/\hbar} \tag{3.13}$$

describes a 7-armed vortex-shaped photoelectron wave packet with a counterclockwise sense of rotation in the polarization plane. The experimental observations of the 7-armed vortex for a time delay of $\tau = -20$ fs is shown in Fig. 3.9c. Although the temporally separated (3ω:4ω) CRCP fields have a circularly symmetric polarization profile (see Fig. 3.3f) the observed vortex-shaped photoelectron wave packet retains the c_7 rotational symmetry. In general, our results show that the symmetry properties of electron wave packets from bichromatic MPI are not fully determined by the laser field symmetry, but completely described by multipath quantum interference of states with different angular momenta.

3.3.5 Rydberg Wave Packet Dynamics

So far, we have discussed different applications of shaper-generated bichromatic fields to sculpt *static* free electron wave packets. An additional perspective of the shaper-based bichromatic scheme is the generation of two-color pump-probe pulse sequences for the observation of ultrafast *dynamics* in atoms and molecules. To this end, one color (pump) is tailored to induce a quantum-dynamical process, i.e. to excite an electronic or nuclear wave packet, while the second color (probe) is variably delayed in time to map the time evolution of the system into the photoionization continuum. Energy- and angle-resolved detection of the resulting photoelectron wave packet provides rich physical information on the process under investigation. By appropriate choice of the two center wavelengths, the pump and the probe step can be disentangled which enables background-free observation of the dynamics. To demonstrate the capabilities of the scheme, we present the results of two time-resolved experiments on prototypical examples of ultrafast electron dynamics. The first experiment is dedicated to the mapping of multiple Rydberg wave packets in the K np- and nf-series [49]. In the second experiment, we observe the spatiotemporal dynamics of a spin-orbit wave packet in the K $4p$ fine structure doublet (see Sect. 3.3.6).

Rydberg atoms are defined as highly-excited quantum systems which are, in general, created by laser pulses driving a ground state electron into Rydberg states with high principal quantum number n close to the ionization threshold [73]. Rydberg series $n\ell$, defined as a set of quantum states with different n but the same angular momentum ℓ, are densely spaced in energy. Sufficiently short i.e. broadband laser pulses therefore excite multiple Rydberg states which form a bound Rydberg wave

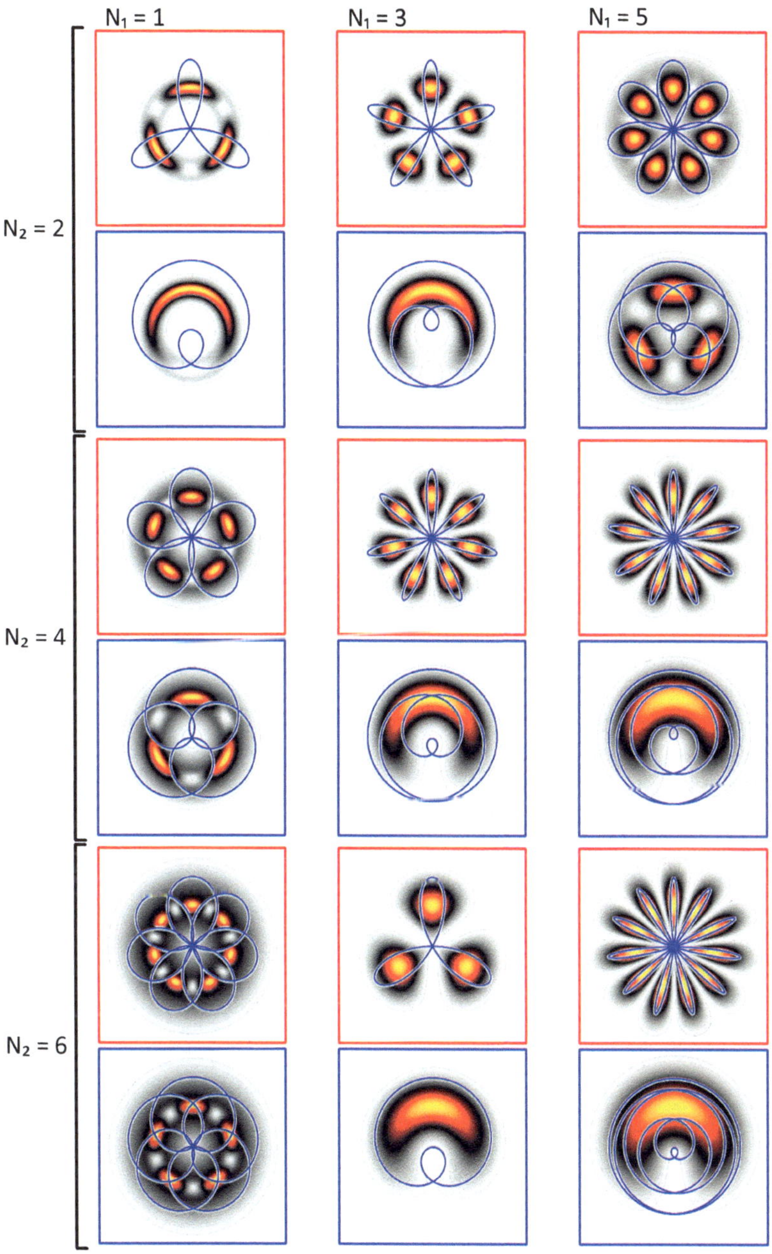

Fig. 3.10 Simulation of photoelectron distributions from coherent N_1- versus N_2-photon ionization using ultrashort CRCP (red frames) or COCP (blue frames) bichromatic fields. To compare the symmetry of the field and the electron distribution the polarization profile of one oscillation period of the bichromatic field (blue curve) is superimposed. For clarity, the electric field is aligned to match the photoelectron wave packets

packet oscillating along the radial coordinate [6, 74–76]. Rydberg wave packets have long been investigated by theory [74, 75] and in experiments [76–79]. In addition, Rydberg atoms are anticipated for future technologies, e.g., as media for data storage and retrieval [80, 81] or as quantum simulators [82, 83].

Here, we employ (3ω:4ω) PLP pulse sequences (cf. Fig. 3.3e) to perform shaper-based two-color pump-probe studies on the dynamics of multiple Rydberg wave packets created by multiphoton excitation of K atoms. The experimental strategy is presented in Fig. 3.11a. The red pulse serves as a pump to create the Rydberg wave packets by three-photon excitation. The blue pulse is time-delayed and serves as a probe to map the dynamics into the continuum via one-photon ionization. The center frequencies of the bichromatic field are chosen to be $\omega_1 \approx 2.125$ rad/fs ($\lambda_1 = 886$ nm) and $\omega_2 \approx \frac{4}{3}\omega_1 = 2.850$ rad/fs ($\lambda_2 = 661$ nm). Both colors have a spectral width of $\Delta\omega_1 = \Delta\omega_2 \approx 0.05$ rad/fs corresponding to a pulse duration of $\Delta t_1 = \Delta t_2 \approx 55$ fs. Linear spectral phase modulation of the blue band with $\varphi_2(\omega) = \tau \cdot (\omega - \omega_2)$ temporally advances ($\tau > 0$) or delays ($\tau < 0$) the blue relatively to the low-frequency (red) pulse, in order to create phase-locked bichromatic pulse sequences for the two-color pump-probe studies. VMI images are measured as a function of τ varied in the interval $\tau = -100, \ldots, 1000$ fs with a step size of $\Delta\tau = 10$ fs.

According to the dipole selection rule $\Delta\ell = \pm 1$ (cf. Sect. 3.3.1), the red pump pulse addresses the np-series via two different three-photon excitation pathways proceeding via the intermediate resonances $5s$ and $3d$. In addition, the nf-series is addressed by a single three-photon pathway via the $3d$-resonance. Therefore, the excited Rydberg wave packets can be described as a superposition of wave functions from the np- and nf-series

$$\psi_{Ry}(t) = \sum_{n}^{N} p_n \, \psi_{np} \, e^{-i\omega_n t} + \sum_{j}^{J} f_j \, \psi_{jf} \, e^{-i\omega_j t}, \tag{3.14}$$

which freely evolve in time with their individual eigenfrequencies ω_n and ω_j, respectively [73, 77, 84]. The population amplitudes p_n and f_j of the Rydberg states are determined by the third order spectrum of the pump pulse which covers the Rydberg states $8f \ldots 24f$ and $9p \ldots 25p$, as shown in the inset to Fig. 3.11a.

One-photon ionization by the time-delayed blue pulse probes the excited Rydberg dynamics into overlapping s-, d- and g-type continua. Measured photoelectron spectra for three selected time delays are displayed exemplarily in Fig. 3.11b–d. Shown are y-z sections through 3D PMDs reconstructed by Abel inversion. For $\tau = -260$ fs (Fig. 3.11b), the blue probe pulse precedes the red pump pulse such that the Rydberg wave packet is created after the probe. Therefore, only the single-color contributions are observed, which overlap at the kinetic energy $\varepsilon_S \approx 1.27$ eV. For $\tau = 0$ (Fig. 3.11c), a two-color photoelectron signal from the excited Rydberg states is observed around $\varepsilon_R = 1.73$ eV. In addition, two signals from fourth-order frequency mixing (cf. Sect. 3.3.1) are observed at distinct energies $\varepsilon_1 = 0.34$ eV and $\varepsilon_2 = 0.81$ eV. Due to the energetic disentanglement of the quantum pathways in

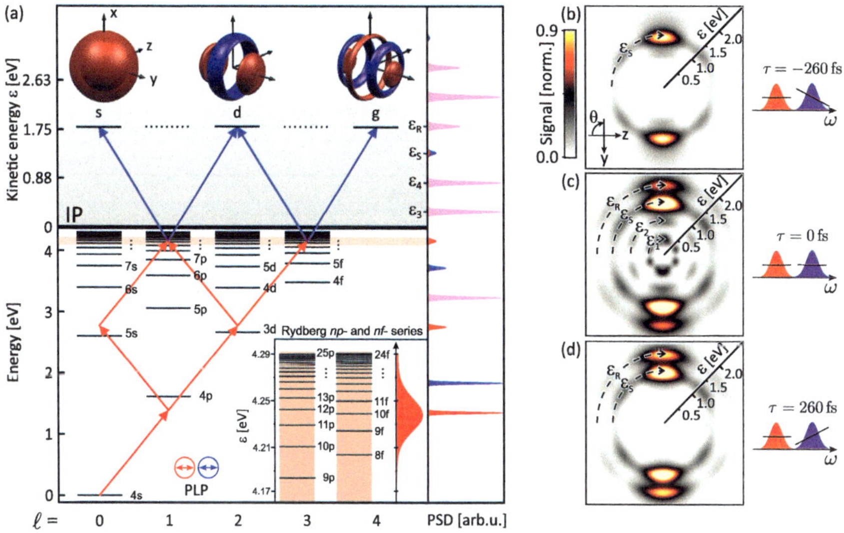

Fig. 3.11 **a** Excitation scheme for K atoms interacting with a (3ω:4ω) PLP pulse sequence. The pump pulse (red arrows) launches Rydberg wave packets in the np- and nf-series via three-photon excitation. Time-delayed one-photon ionization by the probe pulse (blue arrows) maps the Rydberg dynamics onto s-, d- and g-type ionization continua. The corresponding calculated photoelectron wave packets are illustrated on top. The sign of the wave functions is color-coded in red (+) and blue (−). The inset illustrates the two populated Rydberg series covered by the third-order spectrum of the pump pulse, while the right panel displays nonlinear optical spectra up to the fourth order. **b–d** Selected measured and energy-calibrated equatorial y-z-sections through PMDs at **b** $\tau = -260$ fs, **c** $\tau = 0$ fs and **d** $\tau = 260$ fs

bichromatic MPI, the Rydberg signal is clearly separated from both the single-color and the frequency mixing contributions, allowing for background-free detection of the Rydberg dynamics [22]. For $\tau = 260$ fs, the probe follows the pump. Because of the temporal separation of both colors, only the single-color contributions at ε_S and the Rydberg signal at ε_R remain in the spectrum depicted in Fig. 3.11d.

Continuous variation of the time delay reveals rich dynamics of the Rydberg signal in terms of amplitude (yield) and angular distribution. The former results from a modulation of the ionization probability due to the radial oscillation of the two Rydberg wave packets and provides information on their respective compositions. The latter results from a coherent interplay between the interfering s-, d- and g-type photoelectron wave packets, enabling an unambiguous assignment of modes to the underlying Rydberg series. Here, we focus on this angular dynamics of the measured PMDs. To this end, we integrate the recorded VMI images for each time delay τ over the kinetic energy window $\varepsilon \in [1.6, 1.9]$ eV of the Rydberg signal. The resulting angle-resolved photoelectron spectra are plotted in Fig. 3.12a as a function of τ. To highlight small contributions in the angular dynamics observed around $\theta = 180°$, a zoom-in on the interval $\theta \in [120, 240]°$ is displayed in Fig. 3.12b. In both 2D maps, we observe a pronounced signal oscillation in τ-direction with

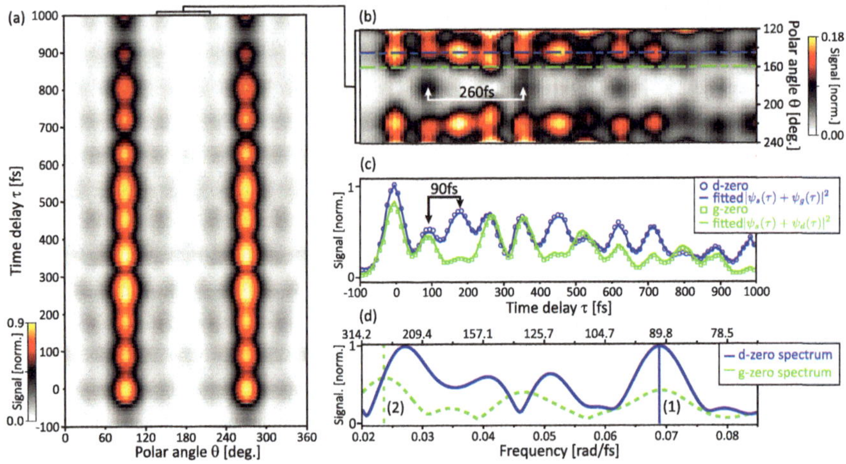

Fig. 3.12 **a** Time- and angle-resolved photoelectron spectra after integration over the energies $\varepsilon_R \approx 1.73\,\mathrm{eV}$ covered by the probed Rydberg wave packets. **b** The zoom-in on the interval $\theta \in [120°, 240°]$ highlights the angular dynamics due to the interference of s-, d- and g-type photoelectron wave packets. **c** Horizontal sections through (**b**) taken along the τ-axis at a g-zero at $\theta = 144°$ (blue circle) and at a d-zero at $\theta = 160°$ (green circles), together with the respective fits (solid lines) [49]. **d** Corresponding Fourier spectra of the experimental data. Two Rydberg modes (1) and (2) discussed in the text are indicated by vertical lines

a period of $T \simeq 90\,\mathrm{fs}$ ($\omega = 0.07\,\mathrm{rad/fs}$). In addition, the spectra show a periodic angular shift which repeats every $T \simeq 260\,\mathrm{fs}$ ($\omega = 0.024\,\mathrm{rad/fs}$), best observable around $\theta = 180°$. For $\tau = 90\,\mathrm{fs}$, we obtain an almost pure g-type angular distribution characterized by two main and three side lobes (first white arrow in Fig. 3.12b). For $\tau = 260\,\mathrm{fs}$, the angular distribution is altered significantly. The central side lobe has vanished due to destructive interference between the different photoelectron wave packets. The remaining two side lobes are slightly shifted towards the equator at $\theta = 180°$.

In the following, we perform an angular-selective analysis to separate the highly entangled Rydberg wave packet dynamics from the Rydberg np- and nf-series. This analysis allows us to uncover the origin of the observed Rydberg modes and, in addition, to assign the observed Rydberg modes to inter- and intra-series beatings between pairs of Rydberg f- and p-states. For this purpose, we extract horizontal sections along the τ-axis from Fig. 3.12b at different polar angles θ, which correspond to the zero-crossings of the d- and g-type wave packets (cf. Fig. 3.12c, d), respectively. This permits us to disentangle the d- and g-type wave packet dynamics up to a background signal originating from s-type wave packets. In Fig. 3.12b, the τ-resolved section extracted at the zero-crossing of the g-type wave packet (g-zero) at $\theta = 160.1°$ (green rectangles) and of the d-type wave packet (d-zero) at $\theta = 144.7°$ (blue circles) are displayed. The respective Fourier spectra are plotted in Fig. 3.12d. Both spectra exhibit a mode labeled by (1) which corresponds to the predominant $T \simeq 90\,\mathrm{fs}$ oscillation mentioned above. Hence, this mode is associated with both an intra-series and an inter-series beating. Using spectroscopic data from [85], the intra-series beating is assigned to states $8f$-$9f$ and the inter-series oscillation occurs between states $11p$-$11f$ and $16p$-$11f$, respectively. In contrast, the mode labeled by (2), corresponding to the $T \simeq 260$ fs oscillation, is exclusively found in the g-zero spectrum. This implies that the slow angular dynamics are mainly caused by contributions from the d-type photoelectron wave packets. Mode (2) is thus identified as an inter-series beating of states $12p$ and $11f$ [85]. To conclude, the pronounced angular dynamics observed in the photoelectron spectrum arise from interference of wave functions with the same parity, i.e. s-, d- and g-type photoelectron wave packets, and are an example for phase control over the differential photoionization cross section [19].

In [49], the angular-selective analysis described above was verified by two independent methods. First, a fitting model (green and blue solid lines in Fig. 3.12c) was employed to decompose the measured photoelectron wave packet dynamics into superpositions of s-, d- and g-type wave packets achieving excellent agreement with the experimental results. Second, a supplementary measurement was performed using COCP pump-probe pulse sequences to selectively address the Rydberg nf-series which is probed exclusively into the g-type continuum. These additional results were in very good accordance with both the differential data analysis and the fitting model.

3.3.6 Spin-Orbit Wave Packet Dynamics

A different class of electron dynamics with distinct physical properties arises due to the coupling of the electron spin and orbital angular momentum. Spin-orbit (SO) interaction gives rise to the fine structure splitting of atomic energy levels into multipletts. In the case of alkali atoms, the $n\ell$ states split into doublets $n\ell_j$ with $j = \ell \pm 1/2$, i.e., into two states separated by the fine structure energy $\Delta\varepsilon$. Coherent excitation of several fine structure states by an ultrashort laser pulse creates a spin-orbit wave packet (SOWP). In contrast to Rydberg wave packets (cf. Sect. 3.3.5) which, in general, oscillate in the radial direction, SO interaction induces an angular motion of the bound electron leading to a realignment of the electron orbital. This angular dynamics can be interpreted as the precession of the spin and orbital angular momentum about the total angular momentum with a precession period given by $T = h/\Delta\varepsilon$ [86].

SOWPs were among the first electron wave packets studied in real time. For example, SOWPs have been observed in Rydberg atoms on the picosecond timescale [87–89] and in alkali atoms [86, 90] and noble gases [91–93] on the femtosecond timescale. Observation of the sub-femtosecond motion of SOWPs in krypton ions was reported in [94]. In addition, coherent control of SOWPs using shaped femtosecond laser pulses was demonstrated [93, 95–97] and the photoionization of atoms by polarization-shaped laser pulses was investigated as a means to produce highly spin-polarized electrons [89, 90, 98, 99].

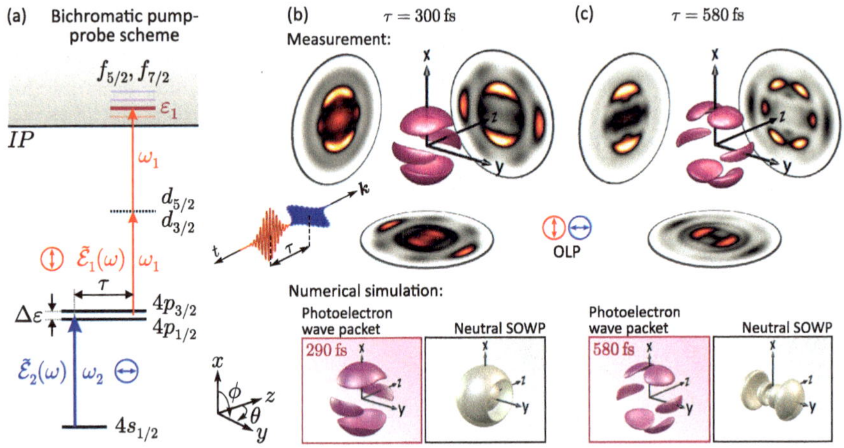

Fig. 3.13 Time-resolved imaging of ultrafast SOWP dynamics in the K $4p$ fine structure doublet using shaper-generated bichromatic OLP pump-probe pulse sequences. The corresponding excitation and ionization scheme is depicted in (**a**). The resonant p-polarized blue pulse (pump) excites the states $4p_{1/2}$ and $4p_{3/2}$ and launches a bound SOWP with an oscillation period of $T = 580$ fs. After a variable time τ the SOWP is probed via non-resonant two-photon ionization by the s-polarized red pulse, mapping the bound state dynamics into the ε_1-channel for background-free detection. For $\tau = 0$, we measure a PMD with quasi c_6 rotational symmetry aligned in the laser polarization plane (see discussion in Sect. 3.3.1; cf. Fig. 3.5c). **b** Around half period $\tau \approx T/2$ the reconstructed PMD rotates by 90° about the x-axis into an alignment coplanar to the laser propagation direction. This rotation indicates the orbital realignment of the SOWP from a dumbbell oriented along the pump polarization (y-axis) into a torus aligned in the x-z-plane (see simulation in bottom insets). **c** After a full cycle $\tau = T$ the measured PMD recovers the initial shape, indicating the return of the SOWP to its dumbbell-shaped spatial distribution

Recently, the incommensurable frequency mixing scheme described in Sect. 3.3.1 was applied in a two-color pump-probe experiment for 3D spatial imaging of ultrafast SOWP dynamics in the K $4p$ fine structure doublet [51]. The corresponding (1+2) REMPI scheme is depicted in Fig. 3.13a. In this experiment, the p-polarized blue pulse serves as a pump to resonantly excite the SOWP composed of the states $4p_{1/2}$ and $4p_{3/2}$. The corresponding oscillation period $T = 580$ fs, derived from the energy splitting of $\Delta\varepsilon = 7$ meV, is much larger than the pulse duration $\Delta t = 80$ fs of both colors. After a variable time delay τ, introduced by linear spectral phase modulation, the SOWP is probed by the s-polarized red pulse via non-resonant two-photon ionization. The probe maps the time evolution of the SOWP exclusively into the ε_1-channel, thus disentangling photoelectron contributions carrying dynamical information from the τ-insensitive single-color contributions at ε_0 and ε_3. By this means, the bichromatic approach provides background-free detection of the bound state dynamics imaged into the 3D PMD of the ε_1-channel. Tomographic reconstruction of the created photoelectron wave packet as a function of time reveals the orbital realignment dynamics of the SOWP in the neutral system.

The starting point of the experiment, $\tau = 0$, was discussed in Sect. 3.3.1 in the context of intra-pulse frequency mixing by temporally overlapping colors. In this case, the photoelectron wave packet in the ε_1-channel is aligned in the laser polarization plane (x-y-plane) and exhibits approximate c_6 rotational symmetry (see p_1-inset to Fig. 3.5). As discussed in [51], this symmetry reflects the initial dumbbell-shaped electron distribution of the excited SOWP, aligned along the y-axis in accordance with the horizontal pump pulse polarization. Next we varied the time delay between the two colors to investigate the time evolution of the SOWP. The results are shown in Fig. 3.13b, c, where we focus on the ε_1-contribution sensitive to the dynamics. At about half the SOWP oscillation period, $\tau \approx T/2$, we observe a counterintuitive rotation of the PMD by 90° about the x-axis. The reconstructed photoelectron density shown in Fig. 3.13b is aligned coplanar to the laser propagation direction in the x-z-plane. In addition, the nodal structure of the photoelectron angular distribution is altered, no longer showing the approximate c_6 symmetry. To interpret the experimental results, we modeled the 3D photoelectron wave packet from MPI of K atoms by bichromatic OLP pulse sequences, taking the electron spin into account. To this end, we numerically solved the time-dependent Schrödinger equation in the coupled basis for the interaction of the neutral atom with the linearly polarized pump and calculated the 3D PMD created by the orthogonally polarized probe using second order time-dependent perturbation theory [51]. The simulated photoelectron density, shown in the bottom left inset to Fig. 3.13b, is in excellent agreement with the measured PMD allowing us to analyze the underlying neutral dynamics. The derived SOWP in the $4p$ fine structure states is illustrated in the right inset. Apparently, after the excitation, the system state has evolved under SO interaction from the initial dumbbell-shaped electron distribution aligned along the y-axis into a torus-shaped distribution aligned in the x-z-plane. This orbital realignment is indicated by the measured photoelectron angular distribution. After a full period, $\tau = T$, the reconstructed photoelectron density recovers the initial shape measured for $\tau = 0$ (cf. Fig. 3.5c). Again we observe the photoelectron wave packet with quasi c_6 rotational symmetry. In accordance with this observation, the simulations yield a dumbbell-shaped neutral electron distribution aligned in pump polarization direction, confirming the completion of a full SOWP cycle.

These results highlight the capabilities of the experimental technique, combining polarization-shaped bichromatic pump-probe sequences with highly differential photoelectron detection, for the background-free observation of spatiotemporal quantum dynamics. A promising future perspective is the use of tailored pump pulses to coherently control and image the time evolution of electron dynamics in atoms and molecules.

3.4 Conclusion and Outlook

In this chapter, we reviewed recent developments in the coherent control of photoelectron momentum distributions (PMDs) using Bichromatic Carrier-Envelope phase-stable Polarization-tailored Supercontinuum (BiCEPS) fields for atomic multiphoton ionization (MPI). For this purpose, continuously tunable BiCEPS fields were generated using a novel white light polarization pulse shaping scheme and a custom composite polarizer [17, 18]. By combining this bichromatic shaping scheme with high-resolution photoelectron tomography based on Velocity Map Imaging (VMI) spectrometry [40], we were able to generate and manipulate three-dimensional (3D) PMDs. To extract the underlying physical mechanisms, we performed our experiments on hydrogen-like atomic model systems (sodium (Na) and potassium (K) atoms). The common theme in the experiments presented in this chapter was the intriguing interplay between the symmetry properties of light fields and the observed PMDs. In six experimental studies, we presented different physical mechanisms for bichromatic MPI. We studied intra-pulse frequency mixing of spectral bands with different ellipticities in coherent control scenarios involving 'N- versus N'-photon and 'N_1- versus N_2'-photon interferences. The analysis of our results showed that the symmetry properties of free electron wave packets are controlled by multipath quantum interference of continuum states with different angular momenta. Depending on the number of photons involved and on the polarization state of the driving fields, different control parameters were active and different symmetries of the free electron wave packets were observed. In the following, we briefly summarize the main results obtained in the experiments presented in Sects. 3.3.1–3.3.6 and discuss some perspectives of our work.

In the first experiment presented in Sect. 3.3.1, we studied bichromatic 'N- versus N'-photon MPI by absorption of photons with different color and polarization leading to high-order intra-pulse frequency mixing [22]. While counter-rotating circularly polarized (CRCP) driving pulses generated four different angular momentum eigenstates, bichromatic orthogonal linearly polarized (OLP) fields gave rise to quasi c_6-rotationally symmetric photoelectron wave packets from third-order intra-pulse frequency mixing and two orthogonal 'f within f'-type wave packets from 3-photon single-color ionization.

In the second study presented in Sect. 3.3.2, we focused on coherent control of the directional photoionization via paths with different numbers of photons ('N_1 versus N_2'). The CEP control of the directionality was achieved by interference of continuum states with opposite parity. Our results showed that shaper-generated bichromatic ($N_1\omega$:$N_2\omega$) fields, with the center frequency ratio tailored to the MPI process, confine the CEP-sensitive asymmetries in a predefined kinetic energy window [23, 24]. In the experiment, (3ω:4ω) parallel linearly polarized (PLP) fields were used to control the MPI of Na atoms. Significant asymmetries in the photoemission were observed along the two distinct directions parallel to the laser polarization. It was demonstrated that the attainable degree of control over the multipath interference in the target state exceeds CEP control by few-cycle ionization scenarios [23].

Full control over the symmetry properties of the 3D PMDs was obtained by using polarization-tailored pulses. To this end, we studied two control scenarios using either single-color or bichromatic circularly polarized pulse sequences as described in Sects. 3.3.3 and 3.3.4. Single color electron vortices, i.e. even-numbered spiral-shaped free electron wave packets from MPI with CRCP single-color pulse sequences, were observed at the ionization threshold and in the above threshold ionization (ATI) [21]. The symmetry properties of the vortices depend on the excitation and ionization dynamics: In the perturbative regime, vortices with c_6 (threshold) and c_8 (ATI) rotational symmetry were observed. In contrast, non-perturbative excitation with a π-pulse led to the formation of an electron vortex with c_4 rotational symmetry. These studies demonstrated the potential of the vortex method for quantum state holography and spectroscopic applications [50].

The use of circularly polarized BiCEPS pulses to create and manipulate photoelectron wave packets with odd-numbered rotational symmetry was reviewed in Sect. 3.3.4. In those experiments, CEP-stable (3ω:4ω) CRCP and COCP laser pulses were employed to generate 7-fold rotationally symmetric and asymmetric crescent-shaped PMDs by MPI of Na atoms [24]. In addition, the creation odd-numbered photoelectron vortices from time-delayed (3ω:4ω) pulse sequences were demonstrated. The interplay between the symmetry properties of the driving field and the resulting electron wave packets was examined by changing optical field parameters such as the relative phases and the CEP. It was demonstrated that the symmetry of the generated wave packets is not fully determined by the symmetry of the laser field, but completely described by multipath quantum interference of states with different angular momenta. In contrast to the even-numbered photoelectron vortices from MPI with CRCP single-color pulse sequences, the odd-numbered photoelectron wave packets from bichromatic MPI were shown to exhibit a non-zero azimuthal probability current and a helical phase front.

To observe ultrafast neutral dynamics in K atoms, we used our pulse shaper to generate continuously tunable and phase-locked two-color pump-probe pulse sequences with variable time delay. In Sects. 3.3.5 and 3.3.6, we demonstrate applications of the shaper-based bichromatic pump-probe scheme on Rydberg wave packet dynamics and on spin-orbit wave packet (SOWP) dynamics. For this purpose, the variable time delay between the two colors of BiCEPS pulse sequence was introduced via linear spectral phase modulation. Due to the energetic disentanglement of the two-color signal from both single-color signals of the pump and the probe pulse, resonant bichromatic frequency mixing was shown to be ideally suited for background-free pump-probe studies of ultrafast quantum dynamics. In the first shaper-based pump-probe scheme presented in Sect. 3.3.5, fourth-order frequency mixing was used for background-free measurement of Rydberg wave packet dynamics [49] with angular and energy resolution. The PMDs from ionization with linearly polarized pump and probe pulses showed pronounced time-, energy- and angle-dependent variations resulting from the dynamics of the bound Rydberg wave packets. In Sect. 3.3.6, we applied the bichromatic pump-probe scheme for time-resolved background-free spatial imaging of ultrafast SOWP dynamics. The analysis revealed a counterintuitive angular precession of the SOWP: As a result of spin-orbit coupling, the initial electron orbital alignment in the laser polarization plane evolved into an orthogonal alignment parallel or coplanar to the laser propagation direction.

Building on our results in coherent control with bichromatic fields, we will extend our studies on the interplay between the symmetry properties of the light field and the observed matter waves towards even more refined control schemes and more complex quantum systems. As shown in Fig. 3.9c, for time delays exceeding the pulse duration, i.e. $\tau > \Delta t$, the symmetry of the PMD deviates significantly from the symmetry of the corresponding driving field. By increasing the intensity, we will study how the control scenarios and the interplay between the symmetries evolve from the multi-photon regime to the tunnel ionization regime. Because our shaping scheme supports modulation of multiple spectral bands, we plan to carry out CEP-sensitive pump-probe studies using multi-color and multi-polarization pulse sequences. Finally, we plan to extend our work on atomic model systems towards small molecules, including chiral species, and investigate bichromatic control of coupled electron nuclear dynamics.

Acknowledgements Financial support by the Deutsche Forschungsgemeinschaft via the DFG Priority Programme SPP 1840 QUTIF is gratefully acknowledged.

References

1. D.J. Tannor, S.A. Rice, Control of selectivity of chemical reaction via control of wave packet evolution. J. Chem. Phys. **83**(10), 5013–5018 (1985)
2. P. Brumer, M. Shapiro, Control of unimolecular reactions using coherent light. Chem. Phys. Lett. **126**(6), 541–564 (1986)
3. H. Rabitz, R. de Vivie-Riedle, M. Motzkus, K. Kompa, Whither the future of controlling quantum phenomena? Science **288**(5467), 824–828 (2000)
4. S.A. Rice, M. Zhao, *Optical Control of Molecular Dynamics* (Wiley, New York, 2000)
5. Y. Silberberg, Quantum coherent control for nonlinear spectroscopy and microscopy. Annu. Rev. Phys. Chem. **60**, 277–292 (2009)
6. K. Ohmori, Wave-packet and coherent control dynamics. Annu. Rev. Phys. Chem. **60**, 487–511 (2009)
7. M. Shapiro, P. Brumer, *Quantum Control of Molecular Processes*, 2nd edn. (Wiley, New York, 2012)
8. M. Wollenhaupt, T. Bayer, T. Baumert, *Control of Ultrafast Electron Dynamics with Shaped Femtosecond Laser Pulses: From Atoms to Solids*, Sect. 4 (Springer, Berlin, 2015), pp. 63–122
9. A.T.J.B. Eppink, D.H. Parker, Velocity map imaging of ions and electrons using electrostatic lenses: application in photoelectron and photofragment ion imaging of molecular oxygen. Rev. Sci. Instrum. **68**(9), 3477–3484 (1997)
10. R. Dörner, V. Mergel, O. Jagutzki, L. Spielberger, J. Ullrich, R. Moshammer, H. Schmidt-Bocking, Cold target recoil ion momentum spectroscopy: a 'momentum microscope' to view atomic collision dynamics. Phys. Rep. **330**(2–3), 95–192 (2000)
11. K. Misawa, Applications of polarization-shaped femtosecond laser pulses. Adv. Phys. X **1**(4), 544–569 (2016)
12. L. Polachek, D. Oron, Y. Silberberg, Full control of the spectral polarization of ultrashort pulses. Opt. Lett. **31**(5), 631–633 (2006)
13. M. Ninck, A. Galler, T. Feurer, T. Brixner, Programmable common-path vector field synthesizer for femtosecond pulses. Opt. Lett. **32**(23), 3379–3381 (2007)
14. D. Kupka, P. Schlup, R.A. Bartels, Simplified ultrafast pulse shaper for tailored polarization states using a birefringent prism. Rev. Sci. Instrum. **80**, 053110 (2009)
15. F. Weise, A. Lindinger, Full control over the electric field using four liquid crystal arrays. Opt. Lett. **34**(8), 1258–12560 (2009)
16. C. Schwarz, O. Huter, T. Brixner, Full vector-field control of ultrashort laser pulses utilizing a single dual-layer spatial light modulator in a common-path setup. JOSA B **32**(5), 933–945 (2015)

17. S. Kerbstadt, L. Englert, T. Bayer, M. Wollenhaupt, Ultrashort polarization-tailored bichromatic fields. J. Mod. Opt. **64**(10–11), 1010 (2017)
18. S. Kerbstadt, D. Timmer, L. Englert, T. Bayer, M. Wollenhaupt, Ultrashort polarization-tailored bichromatic fields from a cep-stable white light supercontinuum. Opt. Express **25**(11), 12518 (2017)
19. M. Shapiro, P. Brumer, *Principles of the Quantum Control of Molecular Processes*, vol. 1 (Wiley, Hoboken, NJ, 2003)
20. M. Wollenhaupt, A. Assion, D. Liese, C. Sarpe-Tudoran, T. Baumert, S. Zamith, M.A. Bouchene, B. Girard, A. Flettner, U. Weichmann, G. Gerber, Interferences of ultrashort free electron wave packets. Phys. Rev. Lett. **89**(17), 173001 (2002)
21. D. Pengel, S. Kerbstadt, D. Johannmeyer, L. Englert, T. Bayer, M. Wollenhaupt, Electron vortices in femtosecond multiphoton ionization. Phys. Rev. Lett. **118**(5), 053003 (2017)
22. S. Kerbstadt, D. Pengel, D. Johannmeyer, L. Englert, T. Bayer, M. Wollenhaupt, Control of photoelectron momentum distributions by bichromatic polarization-shaped laser fields. New J. Phys. **19**, 103017 (2017)
23. S. Kerbstadt, D. Pengel, L. Englert, T. Bayer, M. Wollenhaupt, Carrier-envelope-phase control of asymmetries in the multiphoton ionization of xenon atoms by ultrashort bichromatic fields. Phys. Rev. A **97**(6), 063402 (2018)
24. S. Kerbstadt, K. Eickhoff, T. Bayer, M. Wollenhaupt, Odd electron wave packets from cycloidal ultrashort laser fields. Nat. Commun. **10**(1), 658 (2019)
25. N. Dudovich, T. Polack, A. Peïer, Y. Silberberg, Simple route to strong-field coherent control. Phys. Rev. Lett. **94**, 083002–083002-4 (2005)
26. T. Brixner, G. Gerber, Femtosecond polarization pulse shaping. Opt. Lett. **26**(8), 557–559 (2001)
27. A.M. Weiner, Ultrafast optical pulse shaping: a tutorial review. Opt. Commun. **284**, 3669–3692 (2011)
28. M. Wollenhaupt, M. Krug, J. Köhler, T. Bayer, C. Sarpe-Tudoran, T. Baumert, Photoelectron angular distributions from strong-field coherent electronic excitation. Appl. Phys. B **95**, 245–259 (2009)
29. J. Köhler, M. Wollenhaupt, T. Bayer, C. Sarpe, T. Baumert, Zeptosecond precision pulse shaping. Opt. Express **19**, 11638–11653 (2011)
30. M. Plewicki, F. Weise, S.M. Weber, A. Lindinger, Phase, amplitude, and polarization shaping with a pulse shaper in a mach-zehnder interferometer. Appl. Opt. **45**(32), 8354–8359 (2006)
31. O. Masihzadeh, P. Schlup, R.A. Bartels, Complete polarization state control of ultrafast laser pulses with a single linear spatial light modulator. Opt. Express **15**(26), 18025–18032 (2007)
32. A. Galler, T. Feurer, Pulse shaper assisted short laser pulse characterization. Appl. Phys. B **90**, 427–430 (2008)
33. T. Baumert, T. Brixner, V. Seyfried, M. Strehle, G. Gerber, Femtosecond pulse shaping by an evolutionary algorithm with feedback. Appl. Phys. B **65**(6), 779–782 (1997)
34. D. Yelin, D. Meshulach, Y. Silberberg, Adaptive femtosecond pulse compression. Opt. Lett. **22**(23), 1793–1795 (1997)
35. M. Ivanov, E. Pisanty, High-harmonic generation: taking control of polarization. Nat. Photon. **8**(7), 501–503 (2014)
36. D.B. Milosevic, W. Becker, R. Kopold, Generation of circularly polarized high-order harmonics by two-color coplanar field mixing. Phys. Rev. A **61**(6), 063403 (2000)
37. M. Abel, D.M. Neumark, S.R. Leone, T. Pfeifer, Classical and quantum control of electrons using the carrier-envelope phase of strong laser fields. Laser. Phot. Rev. **5**(3), 352–367 (2011)
38. V. Roudnev, B.D. Esry, General theory of carrier-envelope phase effects. Phys. Rev. Lett. **99**, 220406 (2007)
39. G.A. Garcia, L. Nahon, I. Powis, Two-dimensional charged particle image inversion using a polar basis function expansion. Rev. Sci. Instrum. **75**(11), 4989–4996 (2004)
40. M. Wollenhaupt, M. Krug, J. Köhler, T. Bayer, C. Sarpe-Tudoran, T. Baumert, Three-dimensional tomographic reconstruction of ultrashort free electron wave packets. Appl. Phys. B **95**, 647–651 (2009)

41. C. Smeenk, L. Arissian, A. Staude, D.M. Villeneuve, P.B. Corkum, Momentum space tomographic imaging of photoelectrons. J. Phys. B **42**, 185402 (2009)
42. A.C. Kak, M. Slaney, *Principles of Computerized Tomographic Imaging* (IEEE Press, New York, 1988)
43. J. Itatani, J. Levesque, D. Zeidler, H. Niikura, H. Pepin, J.C. Kieffer, P.B. Corkum, D.M. Villeneuve, Tomographic imaging of molecular orbitals. Nature **432**, 867–871 (2004)
44. K. Bergner, D. Flamm, M. Jenne, M. Kumkar, A. Tünnermann, S. Nolte, Time-resolved tomography of ultrafast laser-matter interaction. Opt. Express **26**(3), 2873–2883 (2018)
45. M. Wollenhaupt, C. Lux, M. Krug, T. Baumert, Tomographic reconstruction of designer free-electron wave packets. Chem. Phys. Chem. **14**, 1341–1349 (2013)
46. C. Lux, M. Wollenhaupt, T. Bolze, Q. Liang, J. Köhler, C. Sarpe, T. Baumert, Circular dichroism in the photoelectron angular distributions of camphor and fenchone from multiphoton ionization with femtosecond laser pulses. Angew. Chem. Int. Ed. **51**, 5001–5005 (2012)
47. C. Lux, M. Wollenhaupt, C. Sarpe, T. Baumert, Photoelectron circular dichroism of bicyclic ketones from multiphoton ionization with femtosecond laser pulses. Chem. Phys. Chem. **16**, 23 (2015)
48. C. Lux, A. Senftleben, C. Sarpe, M. Wollenhaupt, T. Baumert, Photoelectron circular dichroism observed in the above-threshold ionization signal from chiral molecules with femtosecond laser pulses. J. Phys. B **49**(2), 02LT01 (2016)
49. S. Kerbstadt, L. Gabrisch, K. Eickhoff, T. Bayer, M. Wollenhaupt, Imaging multiple rydberg wave packets from shaper-generated two-color femtosecond pump-probe sequences. Phys. Rev. A **99**, 013406 (2019)
50. D. Pengel, S. Kerbstadt, L. Englert, T. Bayer, M. Wollenhaupt, Control of three-dimensional electron vortices from femtosecond multiphoton ionization. Phys. Rev. A **96**, 043426 (2017)
51. T. Bayer, D. Gräfing, S. Kerbstadt, D. Pengel, K. Eickhoff, L. Englert, M. Wollenhaupt, Time-resolved 3d imaging of ultrafast spin-orbit wave packet dynamics. New J. Phys. **21**(3), 033001 (2019)
52. H.A. Bethe, E.E. Salpeter, *Quantum Mechanics of One- and Two-Electron Atoms* (Springer, Berlin, 1957)
53. A.S. Alnaser, I.V. Litvinyuk, Subfemtosecond directional control of chemical processes in molecules. J. Phys. B **50**(3), 032002 (2017)
54. M.F. Kling, J. Rauschenberger, A.J. Verhoef, E. Hasovic, T. Uphues, D.B. Milosevic, H.G. Muller, M.J.J. Vrakking, Imaging of carrier-envelope phase effects in above-threshold ionization with intense few-cycle laser fields. New J. Phys. **10**, 025024–025024-17 (2008)
55. D. Zille, D. Adolph, M. Möller, A.M. Sayler, G.G. Paulus, Chirp and carrier-envelope-phase effects in the multiphoton regime: measurements and analytical modeling of strong-field ionization of sodium. New J. Phys. **20**(6), 063018 (2018)
56. F. Ehlotzky, Atomic phenomena in bichromatic laser fields. Phys. Rep. **345**(4), 176–264 (2001)
57. M.J. Abel, T. Pfeifer, A. Jullien, P.M. Nagel, M.J. Bell, D.M. Neumark, S.R. Leone, Carrier-envelope phase-dependent quantum interferences in multiphoton ionization. J. Phys. B **42**(7), 075601 (2009)
58. A. Wituschek, J. von Vangerow, J. Grzesiak, F. Stienkemeier, M. Mudrich, A simple photoionization scheme for characterizing electron and ion spectrometers. Rev. Sci. Instrum. **87**(8), 083105 (2016)
59. J.M. Ngoko Djiokap, S.X. Hu, L.B. Madsen, N.L. Manakov, A.V. Meremianin, A.F. Starace, Electron vortices in photoionization by circularly polarized attosecond pulses. Phys. Rev. Lett. **115**(11), 113004 (2015)
60. I. Bialynicki-Birula, Z. Bialynicka-Birula, C. Sliwa, Motion of vortex lines in quantum mechanics. Phys. Rev. A **61**(3), 032110 (2000)
61. J.H. Macek, J.B. Sternberg, S.Y. Ovchinnikov, T.G. Lee, D.R. Schultz, Origin, evolution, and imaging of vortices in atomic processes. Phys. Rev. Lett. **102**(14), 143201 (2009)
62. E. Madelung, Quantentheorie in hydrodynamischer form. Z. Phys. **40**(3), 322–326 (1927)
63. L.P.H. Schmidt, C. Goihl, D. Metz, H. Schmidt-Bocking, R. Dörner, S.Y. Ovchinnikov, J.H. Macek, D.R. Schultz, Vortices associated with the wave function of a single electron emitted in slow ion-atom collisions. Phys. Rev. Lett. **112**(8), 083201 (2014)

64. S.M. Lloyd, M. Babiker, G. Thirunavukkarasu, J. Yuan, Electron vortices: beams with orbital angular momentum. Rev. Mod. Phys. **89**(3), 035004 (2017)
65. K.Y. Bliokh, I.P. Ivanov, G. Guzzinati, L. Clark, R. Van Boxem, A. Béché, R. Juchtmans, M.A. Alonso, P. Schattschneider, F. Nori, J. Verbeeck, Theory and applications of free-electron vortex states. Phy. Rep. **690**, 1–70 (2017)
66. K.J. Yuan, H.Z. Lu, A.D. Bandrauk, Photoionization of triatomic molecular ions ${H_3}^{2+}$ by intense bichromatic circularly polarized attosecond UV laser pulses. J. Phys. B **50**(12), 124004 (2017)
67. M. Li, G. Zhang, X. Kong, T. Wang, X. Ding, J. Yao, Dynamic stark induced vortex momentum of hydrogen in circular fields. Opt. Express **26**(2), 878–886 (2018)
68. S. Eckart, M. Richter, M. Kunitski, A. Hartung, J. Rist, K. Henrichs, N. Schlott, H. Kang, T. Bauer, H. Sann, L. Schmidt, M. Schöffler, T. Jahnke, R. Dörner, Nonsequential double ionization by counterrotating circularly polarized two-color laser fields. Phys. Rev. Lett. **117**(13), 133202 (2016)
69. C.A. Mancuso, K.M. Dorney, D.D. Hickstein, J.L. Chaloupka, J.L. Ellis, F.J. Dollar, R. Knut, P. Grychtol, D. Zusin, C. Gentry, M. Gopalakrishnan, H.C. Kapteyn, M.M. Murnane, Controlling nonsequential double ionization in two-color circularly polarized femtosecond laser fields. Phys. Rev. Lett. **117**(13), 133201 (2016)
70. C.A. Mancuso, D.D. Hickstein, K.M. Dorney, J.L. Ellis, E. Hasovic, R. Knut, P. Grychtol, C. Gentry, M. Gopalakrishnan, D. Zusin, F.J. Dollar, X.M. Tong, D.B. Milosevic, W. Becker, H.C. Kapteyn, M.M. Murnane, Controlling electron-ion rescattering in two-color circularly polarized femtosecond laser fields. Phys. Rev. A **93**(5), 053406 (2016)
71. J.M. Ngoko Djiokap, A.V. Meremianin, N.L. Manakov, S.X. Hu, L.B. Madsen, A.F. Starace, Multistart spiral electron vortices in ionization by circularly polarized UV pulses. Phys. Rev. A **94**(1), 013408 (2016)
72. K.J. Yuan, S. Chelkowski, A.D. Bandrauk, Photoelectron momentum distributions of molecules in bichromatic circularly polarized attosecond UV laser fields. Phys. Rev. A **93**(5), 053425 (2016)
73. T.F. Gallagher, *Rydberg Atoms. Cambridge Monographs on Atomic, Molecular* (Cambridge University Press, Cambridge, 2005)
74. G. Alber, P. Zoller, Laser excitation of electronic wave packets in rydberg atoms. Phys. Rep. **199**(5), 231–280 (1991)
75. G. Alber, H. Ritsch, P. Zoller, Generation and detection of rydberg wave packets by short laser pulses. Phys. Rev. A **34**(2), 1058 (1986)
76. L. Noordam, R. Jones, Probing rydberg electron dynamics. J. Mod. Opt. **44**(11–12), 2515–2532 (1997)
77. R.E. Carley, E.D. Boléat, R.S. Minns, R. Patel, H.H. Fielding, Interfering rydberg wave packets in Na. J. Phys. B **38**(12), 1907 (2005)
78. L. Noordam, D. Duncan, T. Gallagher, Ramsey fringes in atomic rydberg wave packets. Phys. Rev. A **45**(7), 4734 (1992)
79. R. Jones, L. Noordam, Electronic wavepackets. Adv. At. Mol. Opt. Phy. **38**, 1–38 (1998)
80. J. Ahn, T.C. Weinacht, P.H. Bucksbaum, Science **287**, 463 (2000)
81. J. Ahn, C. Rangan, D. Hutchinson, P. Bucksbaum, Quantum-state information retrieval in a rydberg-atom data register. Phys. Rev. A **66**(2), 022312 (2002)
82. M. Saffman, Quantum computing with atomic qubits and rydberg interactions: progress and challenges. J. Phys. B: At. Mol. Opt. Phys. **49**(20), 202001 (2016)
83. H. Weimer, M. Müller, I. Lesanovsky, P. Zoller, H.P. Büchler, A rydberg quantum simulator. Nat. Phys. **6**(5), 382 (2010)
84. J. Parker, C.R. Stroud, Coherence and decay of rydberg wave packets. Phys. Rev. Lett. **56**(7), 716–719 (1986)
85. A. Kramida, Y. Rachenko, J. Reader, NIST Atomic Spectra Database (National Institute of Standards and Technology, 2018)
86. S. Zamith, M.A. Bouchene, E. Sokell, C. Nicole, V. Blanchet, B. Girard, Observation of the oscillation of an angular wavepacket. Eur. Phys. J. D **12**, 255–261 (2000)

87. J.F. Christian, L.C. Snoek, S.G. Clement, W.J. van der Zande, Application of associative ionization to the observation of quantum beats in low-lying atomic resonances. Phys. Rev. A **53**(3), 1894–1899 (1996)
88. J.A. Ramswell, V.G. Stavros, J. Lei, Q. Hong, H.H. Fielding, Observation of autoionizing rydberg-electron wave packets in Xe. Phys. Rev. A **59**(3), 2186–2189 (1999)
89. H. Wen, S.N. Pisharody, J.M. Murray, P.H. Bucksbaum, Observing angular precession of a rydberg wave packet due to spin-orbit coupling by orthogonally polarized weak half-cycle pulses. Phys. Rev. A **73**, 052504 (2006)
90. E. Sokell, S. Zamith, M.A. Bouchene, B. Girard, Polarization-dependent pump-probe studies in atomic fine-structure levels: towards the production of spin-polarized electrons. J. Phys. B **33**, 2005–2015 (2000)
91. S. Gilb, E.A. Torres, S.R. Leone, Mapping of time-dependent electron orbital alignment. J. Phys. B **39**(20), 4231 (2006)
92. B. Doughty, L.H. Haber, C. Hackett, S.R. Leone, Photoelectron angular distributions from autoionizing $4s^1 4p^6 6p^1$ states in atomic krypton probed with femtosecond time resolution. J. Chem. Phys. **134**(9), 094307 (2011)
93. J. Plenge, A. Wirsing, C. Raschpichler, B. Wassermann, E. Rühl, Control of coherent excitation of neon in the extreme ultraviolet regime. Faraday Discuss. **153**, 361–373 (2011)
94. E. Goulielmakis, Z.H. Loh, A. Wirth, R. Santra, N. Rohringer, V.S. Yakovlev, S. Zherebtsov, T. Pfeifer, A.M. Azzeer, M.F. Kling, S.R. Leone, F. Krausz, Real-time observation of valence electron motion. Nature **466**, 739–743 (2010)
95. B. Chatel, D. Bigourd, S. Weber, B. Girard, Coherent control of spin-orbit precession with shaped laser pulses. J. Phys. B **41**, 074023–1–074023–5 (2008)
96. S. Zhdanovich, J.W. Hepburn, V. Milner, Strong-field effects in Rabi oscillations between a single state and a superposition of states. Phys. Rev. A **84**(5), 063405 (2011)
97. H. Braun, T. Bayer, D. Pengel, M. Wollenhaupt, T. Baumert, Simultaneous observation of transient and final state dynamics in ultrafast strong-field excitation via time-resolved photoelectron spectroscopy. J. Mod. Opt. **64**(10–11), 1042–1053 (2017)
98. M.A. Bouchene, S. Zamith, B. Girard, Spin-polarized electrons produced by a sequence of two femtosecond pulses. Calculation of differential and global polarization rates. J. Phys. B **34**, 14971512 (2001)
99. T. Nakajima, Control of the spin polarization of photoelectrons/photoions using short laser pulses. Appl. Phys. Lett. **84**(19), 3786–3788 (2004)

Chapter 4
Light-Dressed Spectroscopy of Molecules

Tamás Szidarovszky, Gábor J. Halász, Attila G. Császár, and Ágnes Vibók

Abstract We present a theoretical approach for simulating rovibronic spectra of molecules dressed by medium-intensity laser fields. Numerical results, obtained through the formalism described, are presented for the homonuclear diatomic molecule Na_2, a system suitable for demonstrating and understanding various aspects of light-dressed spectroscopy. We discuss the physical origin of the peaks in the light-dressed spectrum of Na_2 and investigate the light-dressed spectrum in terms of its dependence on the dressing field's intensity and wavelength, temperature, and the turn-on time of the dressing field. The possibility of using light-dressed spectroscopy to derive field-free spectroscopic quantities is also addressed.

4.1 Introduction

Despite the long history of spectroscopy, spanning two centuries, new approaches and methods are being developed in it to this day [1]. This is principally due to the great success of atomic and molecular spectroscopy, both in fundamental research

T. Szidarovszky (✉) · A. G. Császár
Institute of Chemistry, Eötvös Loránd University and MTA-ELTE Complex Chemical Systems Research Group, Pázmány Péter sétány 1/A, Budapest 1117, Hungary
e-mail: tamas821@caesar.elte.hu

A. G. Császár
e-mail: csaszarag@caesar.elte.hu

G. J. Halász
Department of Information Technology, University of Debrecen, PO Box 400, Debrecen 4002, Hungary
e-mail: halasz.gabor@inf.unideb.hu

Á. Vibók
Department of Theoretical Physics, University of Debrecen, PO Box 400, Debrecen 4002, Hungary
e-mail: vibok@phys.unideb.hu

ELI-ALPS, ELI-HU Non-Profit Ltd., Wolfgang Sandner utca 3, Szeged H-6728, Hungary

K. Yamanouchi and D. Charalambidis (eds.), *Progress in Ultrafast Intense Laser Science XV*, Topics in Applied Physics 136,
https://doi.org/10.1007/978-3-030-47098-2_4

and in practical applications, as well as to the remarkable advances in the available experimental techniques and light sources. For example, the development of frequency-comb techniques [2–5] facilitates extremely precise and accurate measurements in the frequency domain, while the development of ultrashort and intense pulsed laser technologies [6, 7] allows for time-resolved spectroscopy on femtosecond or even attosecond [7, 8] timescales.

An often utilized technique of modern spectroscopic methods is the use of two (or more) light pulses. Some of these pulses act as so-called pump pulses, which induce specific changes in the system. Subsequent, so-called probe pulses are used then to measure, directly or indirectly, the changes induced by the pump pulse(s). If the duration of both the pump and the probe pulses is short with respect to the processes investigated, repeating the experiment with varying time delays between the pulses can lead to time-resolved dynamical information [9, 10] or to multidimensional and/or high-resolution spectra [11, 12]. If the pump and probe pulses overlap in time, the signal recorded by the probe pulse does not represent the field-free system, but rather the so-called field-dressed or light-dressed system [13, 14], where the pump pulse acts as the dressing field. If both the pump and the probe pulses are long with respect to the timescales of the processes investigated, one obtains static spectral properties of the light-dressed system. In this chapter we investigate such static spectra of light-dressed systems and henceforth call it *light-dressed spectroscopy*.

For atomic systems the theoretical and experimental methods for investigating optical transitions between light-dressed states is well developed [13]. As to molecules, the concept of light-dressed electronic states, also called light-dressed potentials, has been utilized with considerable success to understand nuclear dynamics both in experimental and theoretical studies [15–23]. To some extent light-dressed rovibrational spectroscopy has been adopted for molecular systems. For example, inducing Autler–Townes-type splittings [24] of rotational transitions with microwave radiation has been used to deduce molecular parameters [25, 26] as well as to promote the spectral assignments of rovibronic levels [27].

In previous theoretical studies the rovibronic spectrum of light-dressed Na_2 was investigated in the context of how the presence of a light-induced conical intersection (LICI) [28, 29], generated by the dressing field, can be identified in the spectrum [30, 31]. In these works all rovibronic degrees of freedom are incorporated into the concept of light-dressed states and the modeling work is carried out for both the dressing field of laser radiation [30] and the quantized dressing field of a microscopic cavity mode [31]. In a separate paper, we provided a more general and detailed discussion on the computation and the properties of light-dressed spectra [32].

In this chapter we review in detail the theory of computing light-dressed spectra induced by laser fields and investigate certain aspects of light-dressed spectroscopy, including its unique properties on deriving spectroscopic information. Our discussion closely follows the one presented in [32].

4.2 Theoretical Approach

Our approach to compute light-dressed spectra of molecules is based on three well-defined steps. First, we compute all field-free molecular rovibronic states relevant to the light-induced processes. Second, using the field-free eigenstates as molecular basis functions, we determine the light-dressed states induced by a medium-intensity light field within the framework of Floquet theory [14, 33, 34]. Third, we compute the transitions between the light-dressed states, induced by a weak probe pulse.

In order to facilitate the required computations, we make the following assumptions: (1) Initially the molecule is in a field-free eigenstate in the gas phase. (2) The molecule is exposed to a medium-intensity dressing light, which is turned on adiabatically, i.e., its envelope varies much slower than the rovibronic timescales characterizing the molecule. (3) The probe pulse, introduced to record the static rovibronic spectrum of the light-dressed molecule, is weak and can be treated in a perturbative manner.

4.2.1 Determination of Light-Dressed States

4.2.1.1 General Considerations

In this section we review some aspects of the Floquet approach [14, 33, 34] which we use to compute the light-dressed states generated by a medium-intensity pump pulse. For a dressed Hamiltonian periodic in time t, such as

$$\hat{H}_{\rm d}(t) = \hat{H}_{\rm mol} + \hat{W}_1(t), \tag{4.1}$$

$$\hat{H}_{\rm d}(t+T) = \hat{H}_{\rm d}(t), \tag{4.2}$$

where $\hat{H}_{\rm mol}$ is the field-free molecular Hamiltonian, $T = 2\pi/\omega_1$ is the time period of the periodicity, and $\hat{W}_1(t)$ is the interaction between the molecule and the dressing field, the time-dependent Schrödinger equation (TDSE)

$$i\hbar\partial_t|\psi(t)\rangle = \hat{H}_{\rm d}(t)|\psi(t)\rangle \tag{4.3}$$

has the general solution

$$|\psi(t)\rangle = \sum_k c_k e^{-\frac{i}{\hbar}\varepsilon_k t}|\Phi_k(t)\rangle, \tag{4.4}$$

where ε_k are the so-called quasienergies, and $|\Phi_k(t)\rangle$ are the Floquet states (also called light-dressed states in our work). The Floquet states satisfy

$$|\Phi_k(t+T)\rangle = |\Phi_k(t)\rangle, \tag{4.5}$$

and

$$(\hat{H}_\mathrm{d}(t) - i\hbar\partial_t)|\Phi_k(t)\rangle = \hat{H}_\mathrm{F}(t)|\Phi_k(t)\rangle = \varepsilon_k|\Phi_k(t)\rangle. \tag{4.6}$$

As can be verified using (4.6), if ε_k is a quasienergy, then $\varepsilon_k + \hbar m\omega_1$ is also a quasienergy with a corresponding Floquet state $e^{im\omega_1 t}|\Phi_k(t)\rangle$, where m is an integer. As (4.4) demonstrates, such a shifted quasienergy does not represent a new physical state, because $\varepsilon_k + \hbar m\omega_1$ with $e^{im\omega_1 t}|\Phi_k(t)\rangle$ gives the same contribution to the wave function as ε_k with $|\Phi_k(t)\rangle$.

Because $|\Phi_k(t)\rangle$ are periodic in time, they can be expanded as a Fourier series,

$$|\Phi_k(t)\rangle = \sum_n |\varphi_{kn}\rangle e^{in\omega_1 t}. \tag{4.7}$$

Substituting (4.7) into (4.6) and assuming $\hat{W}_1(t) = -\mathbf{E}_1\hat{\boldsymbol{\mu}}\cos(\omega_1 t) = -\frac{1}{2}\mathbf{E}_1\hat{\boldsymbol{\mu}}(e^{i\omega_1 t} + e^{-i\omega_1 t})$, (4.6) becomes

$$(\hat{H}_\mathrm{mol} + \hbar n\omega_1)\sum_n |\varphi_{kn}\rangle e^{in\omega_1 t} - \frac{1}{2}\mathbf{E}_1\hat{\boldsymbol{\mu}}\sum_n |\varphi_{kn}\rangle(e^{i(n+1)\omega_1 t} + e^{i(n-1)\omega_1 t})$$
$$= \varepsilon_k \sum_n |\varphi_{kn}\rangle e^{in\omega_1 t}. \tag{4.8}$$

Multiplying (4.8) with $\frac{1}{T}e^{-im\omega_1 t}$ and integrating over the time period T leads to

$$(\hat{H}_\mathrm{mol} + \hbar m\omega_1)|\varphi_{km}\rangle - \frac{1}{2}\mathbf{E}_1\hat{\boldsymbol{\mu}}\big(|\varphi_{k,m-1}\rangle + |\varphi_{k,m+1}\rangle\big) = \varepsilon_k|\varphi_{km}\rangle. \tag{4.9}$$

The Fourier components $|\varphi_{km}\rangle$ can be further expressed as a linear combination of field-free rovibronic molecular eigenstates

$$|\varphi_{km}\rangle = \sum_{\alpha,v,J} C^{(k)}_{m,\alpha v J}|\alpha v J\rangle, \tag{4.10}$$

where α, v, and J represent electronic, vibrational, and rotational quantum numbers, respectively. Using the expansion of (4.10), (4.9) can be turned into the matrix eigenvalue problem

$$\sum_{n,\alpha,v,J} (\mathbf{H}_\mathrm{F})_{m\alpha' v' J', n\alpha v J} C^{(k)}_{n,\alpha v J} = \varepsilon_k C^{(k)}_{m,\alpha' v' J'}, \tag{4.11}$$

where

$$\begin{aligned}(\mathbf{H}_\mathrm{F})_{m\alpha'v'J',n\alpha vJ} \\ &= \left(\langle\alpha'v'J'|\hat{H}_\mathrm{mol}|\alpha vJ\rangle + \hbar m\omega_1\delta_{\alpha\alpha'}\delta_{vv'}\delta_{JJ'}\right)\delta_{nm} \\ &-\frac{1}{2}\langle\alpha'v'J'|\mathbf{E}_1\hat{\boldsymbol{\mu}}|\alpha vJ\rangle\left(\delta_{n,m-1}+\delta_{n,m+1}\right).\end{aligned} \tag{4.12}$$

The pictorial representation of $\mathbf{H}_\mathrm{F}$ reads as

$$\mathbf{H}_\mathrm{F} = \begin{bmatrix} \ddots & \vdots & \vdots & \vdots & \vdots & \vdots & \vdots & \iddots \\ \cdots & \mathbf{H}_\mathrm{A}+\hbar\omega_1\mathbf{I} & \boldsymbol{\lambda} & \mathbf{g}_\mathrm{AA} & \mathbf{g}_\mathrm{AX} & 0 & 0 & \cdots \\ \cdots & \boldsymbol{\lambda}^\dagger & \mathbf{H}_\mathrm{X}+\hbar\omega_1\mathbf{I} & \mathbf{g}_\mathrm{XA} & \mathbf{g}_\mathrm{XX} & 0 & 0 & \cdots \\ \cdots & \mathbf{g}_\mathrm{AA}^\dagger & \mathbf{g}_\mathrm{XA}^\dagger & \mathbf{H}_\mathrm{A} & \boldsymbol{\lambda} & \mathbf{g}_\mathrm{AA} & \mathbf{g}_\mathrm{AX} & \cdots \\ \cdots & \mathbf{g}_\mathrm{AX}^\dagger & \mathbf{g}_\mathrm{XX}^\dagger & \boldsymbol{\lambda}^\dagger & \mathbf{H}_\mathrm{X} & \mathbf{g}_\mathrm{XA} & \mathbf{g}_\mathrm{XX} & \cdots \\ \cdots & 0 & 0 & \mathbf{g}_\mathrm{AA}^\dagger & \mathbf{g}_\mathrm{XA}^\dagger & \mathbf{H}_\mathrm{A}-\hbar\omega_1\mathbf{I} & \boldsymbol{\lambda} & \cdots \\ \cdots & 0 & 0 & \mathbf{g}_\mathrm{AX}^\dagger & \mathbf{g}_\mathrm{XX}^\dagger & \boldsymbol{\lambda}^\dagger & \mathbf{H}_\mathrm{X}-\hbar\omega_1\mathbf{I} & \cdots \\ \iddots & \vdots & \vdots & \vdots & \vdots & \vdots & \vdots & \ddots \end{bmatrix}, \tag{4.13}$$

where each row/column represents a different combination of the values for the α electronic and n Fourier indices of (4.12) (for the sake of simplicity, we assumed only two electronic states, labeled X and A, where X denotes the ground electronic state), and each matrix element in (4.13) is itself a matrix representation of different operators in the space of rovibrational states, i.e.,

$$(\mathbf{H}_\mathrm{A})_{v'J',vJ} = \langle \mathrm{A}v'J'|\hat{H}_\mathrm{mol}|\mathrm{A}vJ\rangle, \tag{4.14}$$

$$(\mathbf{H}_\mathrm{X})_{v'J',vJ} = \langle \mathrm{X}v'J'|\hat{H}_\mathrm{mol}|\mathrm{X}vJ\rangle, \tag{4.15}$$

$$(\boldsymbol{\lambda})_{v'J',vJ} = \langle \mathrm{A}v'J'|\hat{H}_\mathrm{mol}|\mathrm{X}vJ\rangle, \tag{4.16}$$

and

$$(\mathbf{g}_{\alpha\beta})_{v'J',vJ} = -\frac{1}{2}\langle\alpha v'J'|\mathbf{E}_1\hat{\boldsymbol{\mu}}|\beta vJ\rangle, \tag{4.17}$$

and $\mathbf{I}$ is the identity matrix. $\mathbf{H}_\mathrm{A}$ and $\mathbf{H}_\mathrm{X}$ can be thought of as the rovibrational Hamiltonians in the adiabatic electronic states A and X, respectively, while $\boldsymbol{\lambda}$ accounts for intrinsic nonadiabatic couplings between the X and A electronic states. The coupling induced by the dressing field is represented by $\mathbf{g}_{\alpha\beta}$.

After solving (4.11), by diagonalizing the matrix of (4.13), the light-dressed states can be written with the help of (4.7) and (4.10) as

$$|\Phi_k\rangle = \sum_{n,\alpha,v,J} C^{(k)}_{n,\alpha vJ}|\alpha vJ\rangle|n\rangle, \tag{4.18}$$

where we adopt the notation called "*Floquet-state nomenclature*" [33], in which $\langle t|n\rangle = e^{in\omega_1 t}$ and $\omega_1 = 2\pi/T$ with $|\Phi_k(t+T)\rangle = |\Phi_k(t)\rangle$.

4.2.1.2 Simplifying Assumptions

For practical applications (4.13) can often be simplified as follows: (1) If the molecule has no permanent dipole, $\mathbf{g}_{\mathrm{XX}} = \mathbf{g}_{\mathrm{AA}} = \mathbf{0}$. (2) If the intrinsic nonadiabatic couplings can be neglected, $\boldsymbol{\lambda} = \mathbf{0}$. (3) Finally, if $\hbar\omega_1$ is resonant with the electronic excitation between the states X and A, nonresonant light-matter coupling terms can be neglected up to moderate field strengths. This leaves only those $\mathbf{g}_{\alpha\beta}$ matrices nonzero which connect $(\mathbf{H}_{\mathrm{X}} + n\hbar\omega_1\mathbf{I})$-type elements with $(\mathbf{H}_{\mathrm{A}} + (n-1)\hbar\omega_1\mathbf{I})$-type elements. The above three simplifications lead to a block-diagonal $\mathbf{H}_{\mathrm{F}}$, with each two-by-two block being essentially identical, differing only in the value of n in the terms $n\hbar\omega_1\mathbf{I}$ and $(n-1)\hbar\omega_1\mathbf{I}$. The block labeled with the Fourier index n reads

$$\mathbf{H}_{\mathrm{F}}^{2\times 2}(n) = \begin{bmatrix} \mathbf{H}_{\mathrm{X}} + n\hbar\omega_1 & \mathbf{g}_{\mathrm{XA}} \\ \mathbf{g}_{\mathrm{XA}}^{\dagger} & \mathbf{H}_{\mathrm{A}} + (n-1)\hbar\omega_1\mathbf{I} \end{bmatrix}. \tag{4.19}$$

Therefore, instead of solving the general problem presented in (4.11), it becomes sufficient to solve the eigenvalue problem for $\mathbf{H}_{\mathrm{F}}^{2\times 2}(n)$ in order to obtain the light-dressed states and the corresponding quasienergies. Therefore, the above simplifications lead to light-dressed states of the form

$$|\Phi_k(n)\rangle = \sum_{v,J} C_{\mathrm{X}vJ}^{(k)} |\mathrm{X}vJ\rangle|n\rangle + \sum_{v,J} C_{\mathrm{A}vJ}^{(k)} |\mathrm{A}vJ\rangle|n-1\rangle. \tag{4.20}$$

4.2.2 *Temporal Evolution of a Light-Dressed System*

In the representation of (4.12), the $\mathbf{H}_{\mathrm{F}}$ matrix corresponding to the Floquet Hamiltonian and the $C_{n,\alpha vJ}^{(k)}$ expansion coefficients of the Floquet states are both independent of time. Based on this representation, as well as (4.4) and (4.6), the temporal evolution of a light-dressed system can be expressed as [14]

$$\boldsymbol{\Psi}(t) = \sum_k c_k e^{-\frac{i}{\hbar}\mathbf{H}_{\mathrm{F}}t}\boldsymbol{\Phi}_k = e^{-\frac{i}{\hbar}\mathbf{H}_{\mathrm{F}}t}\sum_k c_k\boldsymbol{\Phi}_k = e^{-\frac{i}{\hbar}\mathbf{H}_{\mathrm{F}}t}\boldsymbol{\Psi}(t=0), \tag{4.21}$$

which is formally equivalent to the temporal evolution of a system with a time-independent Hamiltonian.

In a physical scenario when the dressing field amplitude changes slowly with time, those matrix elements of $\mathbf{H}_{\mathrm{F}}$ which represent light-matter couplings change also slowly with time. For a dressing field which is turned on much slower than the characteristic timescales of the field-free system, i.e., adiabatically, (4.21) suggests that the well-known adiabatic theorem could be used to predict the temporal changes in $\mathbf{\Psi}(t)$. Therefore, in the limit of the dressing light intensity going to zero ($\mathbf{g}_{XA} \to 0$), the field-free eigenstates are eigenstates of $\mathbf{H}_F$ as well, and the adiabatic turn-on of the dressing field will convert an initial field-free eigenstate into a single light-dressed state. This means that light-dressed states and field-free states can be correlated in a one-to-one fashion. However, it is important to mention that the one-to-one correlation is not possible if ω_1 is in exact resonance with an allowed transition, because in that case the field-free eigenstates are not eigenstates of $\mathbf{H}_F$ but are linear combinations of $\mathbf{H}_F$ eigenstates even for infinitesimal light-matter coupling strengths. Further information on the temporal evolution of light-dressed states in the Floquet formalism and its possible utilization can be found, for example, in Refs. [35–38].

4.2.3 Transitions Between Light-Dressed States

Once the light-dressed states are determined, one can compute the transition probabilities between the different light-dressed states induced by the weak probe pulse. Following the standard approach of molecular spectroscopy [39], we use first-order time-dependent perturbation theory (TDPT) to derive the relevant equations.

4.2.3.1 General Considerations

Let us assume that the molecule interacts with two periodic electric fields, oscillating with frequencies ω_1 and ω_2. The full Hamiltonian then reads as

$$\hat{H}(t) = \hat{H}_{\text{mol}} + \hat{W}_1(t) + \hat{W}_2(t), \tag{4.22}$$

where $\hat{W}_1(t)$ and $\hat{W}_2(t)$ represent the interaction between the molecule and the two fields. In the dipole approximation

$$\begin{aligned} \hat{W}_1(t) &= -\mathbf{E}_1\hat{\boldsymbol{\mu}}\cos(\omega_1 t + \phi) \\ \hat{W}_2(t) &= -\mathbf{E}_2\hat{\boldsymbol{\mu}}\cos(\omega_2 t). \end{aligned} \tag{4.23}$$

$\hat{W}_1(t)$ generates the light-dressed states, while $\hat{W}_2(t)$ represents the interaction with a weak probe pulse used to record the spectrum of the light-dressed molecule. The formation of light-dressed states by $\hat{W}_1(t)$ is described within the Floquet approach, as overviewed in Sect. 4.2.1. For computing the $\hat{W}_2(t)$-induced transition amplitudes between individual light-dressed states or between the superpositions of light-dressed states, first-order time-dependent perturbation theory is used. To derive our working equations, we start with the TDSE containing the interaction with both the dressing and the probe fields,

$$i\hbar\partial_t|\Psi(t)\rangle = (\hat{H}_{\text{mol}} + \hat{W}_1(t) + \hat{W}_2(t))|\Psi(t)\rangle = (\hat{H}_{\text{d}}(t) + \hat{W}_2(t))|\Psi(t)\rangle. \quad (4.24)$$

Equation (4.24) is transformed to the interaction picture using the transformation

$$|\Psi_{\text{I}}(t)\rangle = e^{\frac{i}{\hbar}\int\limits_{t_0}^{t}\hat{H}_{\text{d}}(t')dt'}|\Psi(t)\rangle, \quad (4.25)$$

which leads to

$$i\hbar\partial_t|\Psi_{\text{I}}(t)\rangle = \hat{W}_{2\text{I}}(t)|\Psi_{\text{I}}(t)\rangle, \quad (4.26)$$

where $\hat{W}_{2\text{I}}(t) = e^{\frac{i}{\hbar}\int_{t_0}^{t}\hat{H}_{\text{d}}(t')dt'}\hat{W}_2(t)e^{-\frac{i}{\hbar}\int_{t_0}^{t}\hat{H}_{\text{d}}(t')dt'}$. Following the usual TDPT procedure of integrating (4.26) from t_0 to t and applying successive approximations to express $|\Psi_{\text{I}}(t)\rangle$ gives

$$|\Psi_{\text{I}}(t)\rangle = \hat{U}(t, t_0)|\Psi_{\text{I}}(t_0)\rangle \quad (4.27)$$

with

$$\hat{U}(t, t_0) = \hat{I} + \frac{1}{i\hbar}\int\limits_{t_0}^{t}\hat{W}_{2\text{I}}(t')dt' + \frac{1}{(i\hbar)^2}\int\limits_{t_0}^{t}\hat{W}_{2\text{I}}(t')\int\limits_{t_0}^{t'}\hat{W}_{2\text{I}}(t'')dt''dt' + \cdots. \quad (4.28)$$

Taking the first two terms of the propagator in (4.28) leads to

$$|\Psi_{\text{I}}(t)\rangle = |\Psi_{\text{I}}(t_0)\rangle + \frac{1}{i\hbar}\int\limits_{t_0}^{t}\hat{W}_{2\text{I}}(t')dt'|\Psi_{\text{I}}(t_0)\rangle. \quad (4.29)$$

The amplitude of the transition to a final state $|\Psi_{\text{I}}^{(\text{F})}\rangle$ at time t is thus

$$\langle\Psi_{\text{I}}^{(\text{F})}|\Psi_{\text{I}}(t)\rangle = \langle\Psi_{\text{I}}^{(\text{F})}|\Psi_{\text{I}}(t_0)\rangle + \frac{1}{i\hbar}\int\limits_{t_0}^{t}\langle\Psi_{\text{I}}^{(\text{F})}|\hat{W}_{2\text{I}}(t')|\Psi_{\text{I}}(t_0)\rangle dt'. \quad (4.30)$$

By expanding $|\Psi_{\text{I}}^{(\text{F})}\rangle$ and $|\Psi_{\text{I}}(t_0)\rangle$ as a superposition of the $|\Phi_k(t_0)\rangle$ Floquet states,

$$|\Psi_{\text{I}}^{(\text{F})}\rangle = \sum_l a_l e^{-\frac{i}{\hbar}\varepsilon_l t_0}|\Phi_l(t_0)\rangle \quad (4.31)$$

and

$$|\Psi_{\text{I}}(t_0)\rangle = \sum_k b_k e^{-\frac{i}{\hbar}\varepsilon_k t_0}|\Phi_k(t_0)\rangle, \quad (4.32)$$

equation (4.30) gives

$$\langle\Psi_{\mathrm{I}}^{(\mathrm{F})}|\Psi_{\mathrm{I}}(t)\rangle = \sum_{l,k} a_l^* b_k e^{-\frac{i}{\hbar}(\varepsilon_k-\varepsilon_l)t_0}\langle\Phi_l(t_0)|\Phi_k(t_0)\rangle$$
$$+\frac{1}{i\hbar}\sum_{l,k} a_l^* b_k \int_{t_0}^{t} e^{-\frac{i}{\hbar}(\varepsilon_k-\varepsilon_l)t_0}\langle\Phi_l(t_0)|\hat{W}_{2\mathrm{I}}(t')|\Phi_k(t_0)\rangle dt'$$
$$= \sum_{l,k} a_l^* b_k e^{-\frac{i}{\hbar}(\varepsilon_k-\varepsilon_l)t_0}\langle\Phi_l(t_0)|\Phi_k(t_0)\rangle$$
$$+\frac{1}{i\hbar}\sum_{l,k} a_l^* b_k \int_{t_0}^{t} e^{-\frac{i}{\hbar}(\varepsilon_k-\varepsilon_l)t_0}\langle\Phi_l(t_0)|e^{\frac{i}{\hbar}\int_{t_0}^{t'}\hat{H}_{\mathrm{d}}(t'')dt''}\hat{W}_2(t')e^{-\frac{i}{\hbar}\int_{t_0}^{t'}\hat{H}_{\mathrm{d}}(t'')dt''}|\Phi_k(t_0)\rangle dt'. \quad (4.33)$$

Because $e^{-\frac{i}{\hbar}\varepsilon_k t}|\Phi_k(t)\rangle$ is a solution of the TDSE of (4.3), the effect of the $\hat{U}_{\mathrm{d}}(t', t_0) = e^{-\frac{i}{\hbar}\int_{t_0}^{t'}\hat{H}_{\mathrm{d}}(t'')dt''}$ operator, describing $\hat{H}_{\mathrm{d}}(t)$-governed time evolution from t_0 to t', can be evaluated as $\hat{U}_{\mathrm{d}}(t', t_0)(e^{-\frac{i}{\hbar}\varepsilon_k t_0}|\Phi_k(t_0)\rangle) = e^{-\frac{i}{\hbar}\varepsilon_k t'}|\Phi_k(t')\rangle$. Utilizing this expression, one obtains

$$\langle\Psi_{\mathrm{I}}^{(\mathrm{F})}|\Psi_{\mathrm{I}}(t)\rangle = \sum_{l,k} a_l^* b_k e^{-\frac{i}{\hbar}(\varepsilon_k-\varepsilon_l)t_0}\langle\Phi_l(t_0)|\Phi_k(t_0)\rangle$$
$$+\frac{1}{i\hbar}\sum_{l,k} a_l^* b_k \int_{t_0}^{t}\langle\Phi_l(t')|\hat{W}_2(t')|\Phi_k(t')\rangle e^{-\frac{i}{\hbar}(\varepsilon_k-\varepsilon_l)t'}dt'. \quad (4.34)$$

Finally, using the explicit form of $\hat{W}_2(t)$ given in (4.23) and expressing the periodic $|\Phi_k(t)\rangle$ functions under the integral with their Fourier series [see (4.7)], we obtain

$$\langle\Psi_{\mathrm{I}}^{(\mathrm{F})}|\Psi_{\mathrm{I}}(t)\rangle = \sum_{l,k} a_l^* b_k e^{-\frac{i}{\hbar}(\varepsilon_k-\varepsilon_l)t_0}\langle\Phi_l(t_0)|\Phi_k(t_0)\rangle$$
$$-\frac{1}{2i\hbar}\sum_{l,k} a_l^* b_k \sum_{n,m}\int_{t_0}^{t}\langle\varphi_{ln}|\mathbf{E}_2\hat{\boldsymbol{\mu}}|\varphi_{km}\rangle e^{-\frac{i}{\hbar}(\hbar\omega_1(n-m)+\varepsilon_k-\varepsilon_l\pm\hbar\omega_2)t'}dt'. \quad (4.35)$$

4.2.3.2 Simplifying Assumptions

For molecules with negligible intrinsic nonadiabatic couplings and no or negligibly small permanent dipole, such as the Na_2 molecule investigated below, the light-dressed states determined within a Floquet approach, in which nonresonant coupling terms with the dressing field are neglected [see (4.19)], (4.7) can be written as

$$|\Phi_k(t)\rangle = \sum_n (|\alpha_{kn}\rangle + |\beta_{k(n-1)}\rangle e^{-i\omega_1 t})e^{in\omega_1 t}, \quad (4.36)$$

where $|\alpha_{kn}\rangle$ and $|\beta_{k(n-1)}\rangle$ represent the two manifolds of rovibronic states with Fourier indices n and $n-1$, respectively. If the nonresonant coupling terms are neglected in the Floquet approach, the Floquet Hamiltonian becomes block diagonal [see (4.19)], $|\alpha_{kn}\rangle$ and $|\beta_{k(n-1)}\rangle$ become the same for all n, and (4.36) is simplified to

$$|\Phi_k(t)\rangle = (|\alpha_k\rangle + |\beta_k\rangle e^{-i\omega_1 t})e^{in\omega_1 t}. \tag{4.37}$$

In (4.37), $|\Phi_k(t)\rangle$ is a Floquet state obtained from the nth two-by-two block of the Floquet Hamiltonian, and the quasienergy associated with $|\Phi_k(t)\rangle$ may be written as $\varepsilon_k + \hbar n\omega_1$. As explained under (4.6), if the quasienergy is shifted by $-\hbar n\omega_1$ and the Floquet state is multiplied by $e^{-in\omega_1 t}$, one obtains an equivalent physical state. Therefore, (4.37) can be rewritten as

$$|\Phi_k(t)\rangle = |\alpha_k\rangle + |\beta_k\rangle e^{-i\omega_1 t}, \tag{4.38}$$

with the corresponding quasienergy ε_k. Using (4.38) instead of (4.7) leads to a simplified version of (4.35),

$$\begin{aligned}\langle\Psi_{\mathrm{I}}^{(\mathrm{F})}|\Psi_{\mathrm{I}}(t)\rangle &= \sum_{l,k} a_l^* b_k e^{-\frac{i}{\hbar}(\varepsilon_k-\varepsilon_l)t_0}\langle\Phi_l(t_0)|\Phi_k(t_0)\rangle \\ &\quad -\frac{1}{2i\hbar}\sum_{l,k} a_l^* b_k \int_{t_0}^{t}\langle\beta_l|\mathbf{E}_2\hat{\boldsymbol{\mu}}|\alpha_k\rangle e^{-\frac{i}{\hbar}(\varepsilon_k-\varepsilon_l-\hbar\omega_1\pm\hbar\omega_2)t'}dt' \\ &\quad -\frac{1}{2i\hbar}\sum_{l,k} a_l^* b_k \int_{t_0}^{t}\langle\alpha_l|\mathbf{E}_2\hat{\boldsymbol{\mu}}|\beta_k\rangle e^{-\frac{i}{\hbar}(\varepsilon_k-\varepsilon_l+\hbar\omega_1\pm\hbar\omega_2)t'}dt', \end{aligned} \tag{4.39}$$

where we use the fact that $\langle\alpha_l|\mathbf{E}_2\hat{\boldsymbol{\mu}}|\alpha_k\rangle = \langle\beta_l|\mathbf{E}_2\hat{\boldsymbol{\mu}}|\beta_k\rangle = 0$ for a molecule with no permanent dipole. By following the standard TDPT procedure, and assuming that the initial and final states of the transition in (4.39) are the kth and lth light-dressed states, the $T_{l\leftarrow k}$ transition amplitude becomes

$$\begin{aligned} T_{l\leftarrow k} &\propto \sum_{v,J}\sum_{v',J'} C^{(l)*}_{\mathrm{A}v'J'} C^{(k)}_{\mathrm{X}vJ}\langle \mathrm{A}v'J'|\mathbf{E}_2\hat{\boldsymbol{\mu}}|\mathrm{X}vJ\rangle\delta(\varepsilon_k-\varepsilon_l-\hbar\omega_1\pm\hbar\omega_2) \\ &+\sum_{v,J}\sum_{v',J'} C^{(l)*}_{\mathrm{X}v'J'} C^{(k)}_{\mathrm{A}vJ}\langle \mathrm{X}v'J'|\mathbf{E}_2\hat{\boldsymbol{\mu}}|\mathrm{A}vJ\rangle\delta(\varepsilon_k-\varepsilon_l+\hbar\omega_1\pm\hbar\omega_2). \end{aligned} \tag{4.40}$$

Based on the arguments of the delta functions in (4.40), the first term can be interpreted as a transition between the light-dressed states having quasienergies ε_k and $\varepsilon_l + \hbar\omega_1$, that is, in the notation of (4.20), a transition between $|\Phi_k(n)\rangle$ and $|\Phi_l(n')\rangle$ with $n = n' - 1$. Similarly, the second term in (4.40) can be interpreted as a transition between $|\Phi_k(n)\rangle$ and $|\Phi_l(n')\rangle$ with $n = n' + 1$.

4.3 Computational Details

To illustrate the numerical results one can derive *via* the theory introduced in Sect. 4.2, we investigate the Na_2 molecule. In the simulations we consider the $X\,^1\Sigma_g^+$ ground state and the first excited $A\,^1\Sigma_u^+$ electronic state of Na_2, which we represent with the potential energy curves (PEC) on [40]. The transition dipole function between these two states is taken from [41]. The field-free rovibrational wave functions of Na_2 on the $V_X(R)$ and $V_A(R)$ PECs are computed using 200 spherical-DVR basis functions [42] with the related grid points placed in the internuclear coordinate range (0, 10) bohr. All rovibrational eigenstates with $J < 16$ and whose energy does not exceed the zero-point energy of the respective PEC by more than $2000\,\text{cm}^{-1}$ were included in the basis representing the Floquet Hamiltonian of (4.19), which was then diagonalized to obtain the light-dressed states.

In all computations the polarization vector of the pump pulse was assumed to be parallel to the polarization vector of the probe pulse. Therefore, the projection of the total angular momentum onto this axis is a conserved quantity in our simulations.

When investigating the effect of the turn-on time of the dressing field (see Sect. 4.4.5), the TDSE was solved using the simple formula $\boldsymbol{\Psi}(t+dt) = e^{-(i/\hbar)\mathbf{H}(t)dt}\boldsymbol{\Psi}(t)$. Due to the small size of $\mathbf{H}(t)$ (few thousand by few thousand) the exponential function could be constructed by diagonalizing $\mathbf{H}(t)$ at each time step.

4.4 Results and Discussion

4.4.1 Interpretation of the Light-Dressed Spectrum

Before the light-dressed spectra are investigated in detail, it is worth considering their expected structure qualitatively. Naturally, the light-dressed spectrum strongly depends on the molecule investigated and the properties of the dressing field. For Na_2 the rotational, vibrational, and electronic transition wavenumbers considered in this work are of the orders of 1,100, and $15{,}000\,\text{cm}^{-1}$, respectively. Figure 4.1 depicts the landscape of light-dressed PECs for the Na_2 molecule dressed by light whose wavelength is $\lambda = 662$ nm. As can be seen in Fig. 4.1, the manifolds of light-dressed states labeled by n are well separated from each other. Based on (4.40), arrows are drawn to indicate the physical origin of possible absorption and stimulated emission processes induced by the probe pulse. Absorption is described by the first term in (4.40), in which the initial light-dressed state contributes through its X ground electronic state component and the final state contributes through its A excited electronic state component. On the other hand, stimulated emission originates from the second term in (4.40), in which the initial light-dressed state contributes through its A excited electronic state component and the final state contributes through its X ground electronic state component.

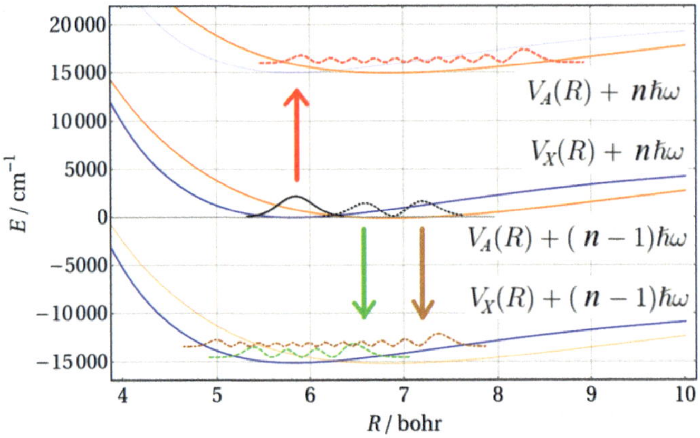

Fig. 4.1 Light-dressed diabatic potential-energy curves (PECs) of Na_2 obtained with a dressing-light whose wavelength is $\lambda = 662$ nm. The energy scale stands for the quasienergy of Floquet theory [33]. Probability densities from the vibrational wave functions are drawn for the $|X\ 0\ 0\rangle|n\rangle$ (continous black line on the $V_X(R) + n\hbar\omega_1$ PEC), $|X\ 3\ 0\rangle|n-1\rangle$ (green dashed line on the $V_X(R) + (n-1)\hbar\omega_1$ PEC), $|X\ 11\ 0\rangle|n-1\rangle$ (brown dashed line on the $V_X(R) + (n-1)\hbar\omega_1$ PEC), $|A\ 1\ 1\rangle|n-1\rangle$ (black dotted line on the $V_A(R) + (n-1)\hbar\omega_1$ PEC), and $|A\ 9\ 1\rangle|n\rangle$ (red dashed line on the $V_A(R) + n\hbar\omega_1$ PEC) states. Up- and downward vertical arrows represent absorption and stimulated emission, respectively. The two product states with the largest contribution to the light-dressed state correlating to $|X\ 0\ 0\rangle$ at $|E_1| \to 0$ are $|X\ 0\ 0\rangle|n\rangle$ and $|A\ 1\ 1\rangle|n-1\rangle$

4.4.2 Intensity Dependence of the Light-Dressed Spectrum

Figure 4.2 shows absorption and stimulated emission spectra of Na_2 at 0 K, when dressed by light fields of different intensity and a wavelength of 662 nm.

As seen in Fig. 4.2, the height of the envelopes of both the absorption and the stimulated emission spectra increase with increasing dressing-field intensity, while the stimulated emission peaks disappear at the limit of zero dressing-field intensity. We point out here that the light-dressed states and the corresponding light-dressed spectra change when the dressing field wavelength is changed. Therefore, if dressing light wavelengths different from 662 nm are used, the height of the envelope of the absorption spectrum might decrease or even fluctuate with increasing dressing light intensity.

Inspecting the individual transition lines reveals that the light-dressing process leads to the splitting of field-free absorption peaks as well as the appearance of new peaks, as shown in the upper and lower panels of Fig. 4.3, respectively. Assignment and understanding of the physical origin of the transition peaks of the light-dressed spectrum can be based on the selection rules governing transitions between field-free states and on the fact that the light-dressed states can be described as a superposition of field-free states.

For example, the upper panel of Fig. 4.3 shows the progression of three peaks, corresponding to the transitions from the initial state (the light-dressed state corre-

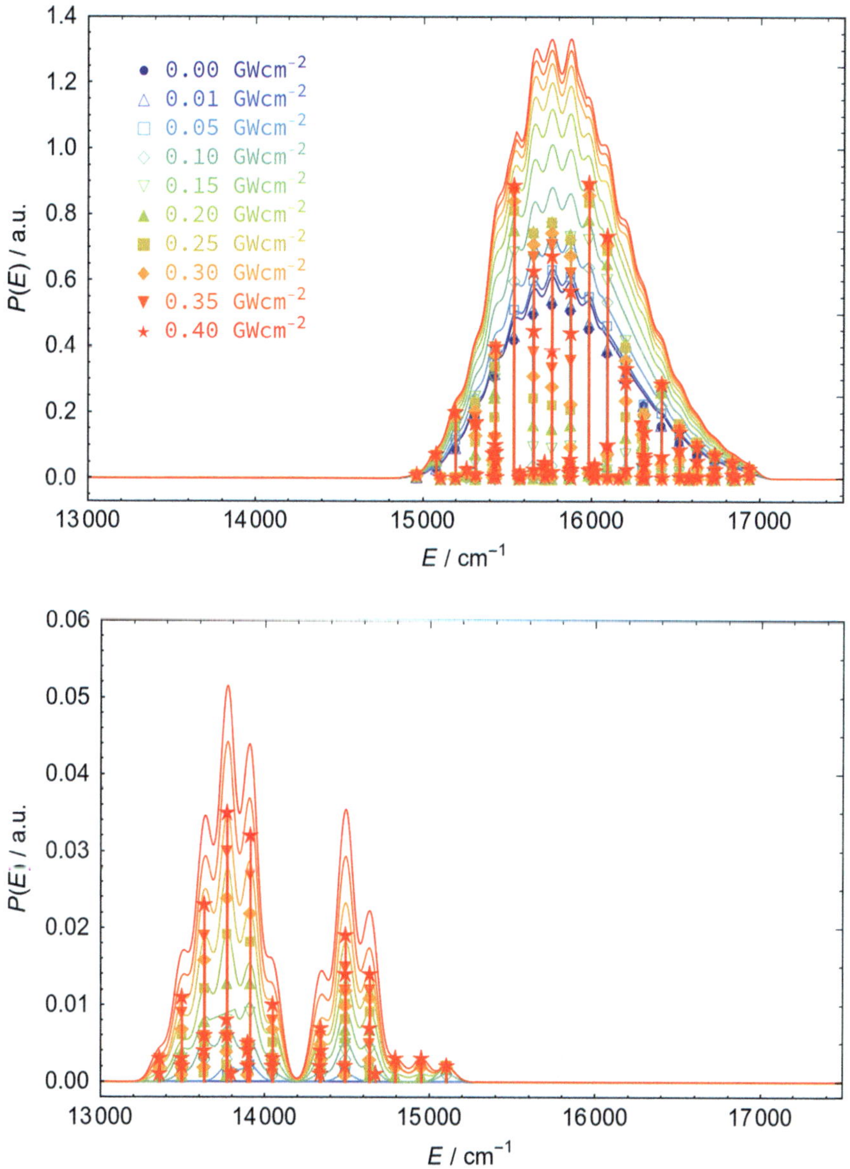

Fig. 4.2 Absorption (upper panel) and stimulated emission (lower panel) spectra of Na_2 dressed with 662 nm wavelength laser lights of different intensity, at 0 K. The stick spectra were computed using (4.40) and show transitions from the field-dressed state, which correlates to the $|X\ 0\ 0\rangle$ rovibronic ground state in the limit of the dressing field intensity going to zero. The envelopes shown are obtained by taking the convolution of the stick spectra with a Gaussian function having a standard deviation of $50\,\text{cm}^{-1}$

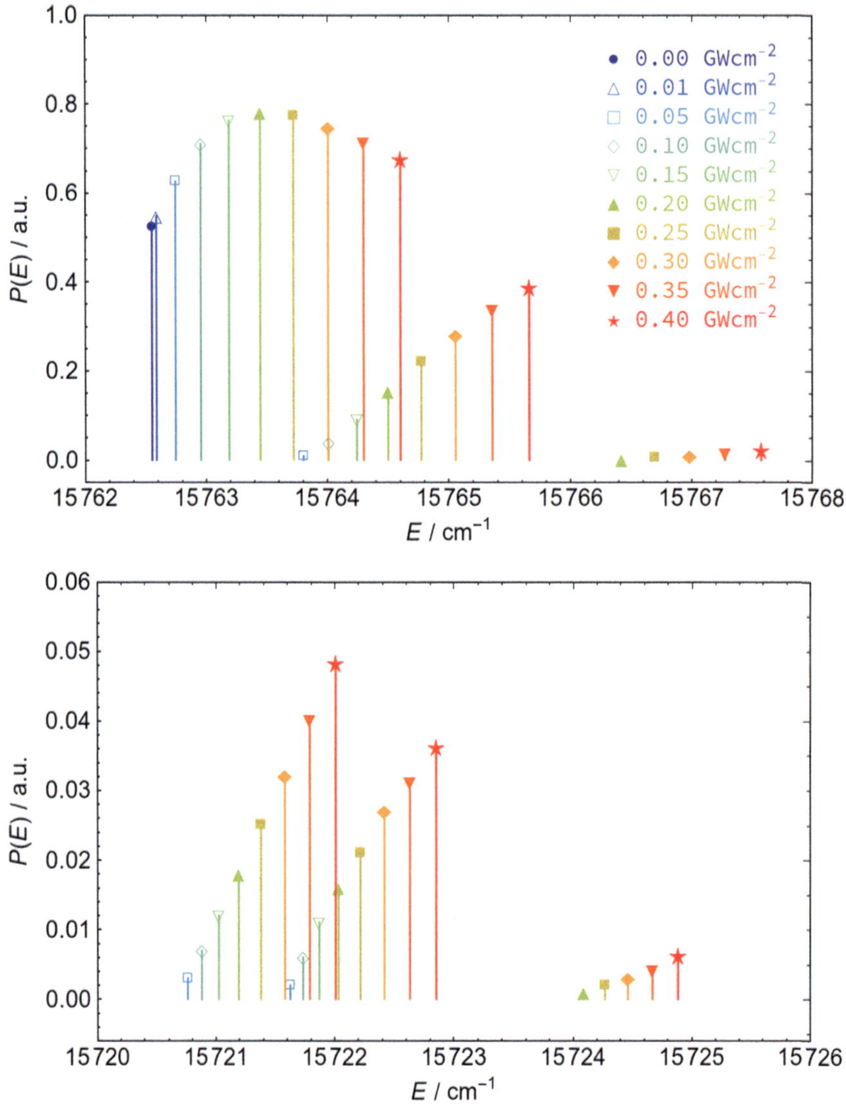

Fig. 4.3 Progression of selected light-dressed absorption peaks of Na_2 dressed with a 662 nm wavelength laser light at 0 K. The upper panel shows peaks originating from the field-free transition $|X\ 0\ 0\rangle \rightarrow |A\ 7\ 1\rangle$, while the lower panel shows peaks which have no field-free counterpart

lating to the field-free ground state) composed primarily of the $|X\ 0\ 0\rangle$ state with smaller contributions from the $|X\ 0\ 2\rangle$ and $|A\ 1\ 1\rangle$ states to the light-dressed states composed primarily of the $|A\ 7\ 1\rangle$, $|A\ 7\ 3\rangle$, and $|A\ 7\ 5\rangle$ states, with $|X\ v\ J\rangle$-type states (J even) contributing as well. These transitions can be interpreted as those originating from the field-free transition $|X\ 0\ 0\rangle \rightarrow |A\ 7\ 1\rangle$, which is split due to the mixing of field-free states through the light-matter coupling with the dressing field. Such type of peak splittings are similar to the well-known *Autler–Townes effect* [24], which is a useful tool in spectroscopy, see, for example [25–27]. The upper panel of Fig. 4.3 not only demonstrates splitting of the peak, in which the sum of individual peak intensities remain unchanged, but exhibits an overall increase in the peak intensities when the strength of the dressing field is increased. In spectroscopy the changes in transition peak intensities resulting from couplings between eigenstates of a zeroth-order Hamiltonian is called *intensity borrowing* [39].

The lower panel of Fig. 4.3 shows the progression of three peaks, which appear as new peaks rather than arising from the splitting of a field-free peak. These transitions occur between the initial state (light dressed state correlating to the field-free ground state) composed primarily of $|X\,0\,0\rangle$ with smaller contributions from $|X\,0\,2\rangle$ and $|A\,1\,1\rangle$ and the light-dressed states composed primarily of the $|X\,4\,0\rangle$, $|X\,4\,2\rangle$, and $|X\,4\,4\rangle$ states. Such transitions are forbidden in the limit of zero dressing-light intensity; however, they become visible as the light-matter coupling with the dressing field mixes the $|X\,4\,J\rangle$ states (J even) with $|A\,v\,1\rangle$-type states, to which $|X\,0\,0\rangle$ has non-zero transition probability. The appearance of transition peaks as a result of such a mixing can be understood as an intensity-borrowing effect.

As to the stimulated emission peaks shown in the lower panel of Fig. 4.2, they represent transitions from the initial state to the light-dressed states composed primarily of the vibrationally highly excited $|X\,v\,0\rangle$- and $|X\,v\,2\rangle$-type states, with $|A\,v'\,J\rangle$-type states (J odd) also giving a small contribution.

4.4.2.1 Predicting Field-Free Properties Using Extrapolation

Although light-dressed spectroscopy might yield transition peaks forbidden in the field-free case, the transition wavenumbers between light-dressed states are in general different from the transition wavenumbers between field-free states. If one is interested in obtaining field-free transition wavenumbers, one can record the light-dressed spectrum at several dressing-field intensities and extrapolate to the zero intensity limit. Such a procedure is of course most valuable when the transition is forbidden in the field-free case.

As an example, we examine the stimulated emission peak at around 13,770 cm^{-1} (see Fig. 4.4). The emission peak around 13,770 cm^{-1} represents a transition in which the initial state is composed primarily of the $|X\,0\,0\rangle$ ground state and has small contributions from the $|X\,0\,2\rangle$ and $|A\,1\,1\rangle$ states, and in which the final state is composed primarily of $|X\,9\,2\rangle$, with $|X\,9\,0\rangle$ and $|A\,13\,1\rangle$ giving a small contribution, as well. In the limit of zero dressing-light intensity, the initial and final states correlate to $|X\,0\,0\rangle$ and $|X\,9\,2\rangle$, respectively. Transitions between these two field-free states is forbidden; nonetheless, their accurate transition wavenumber can be obtained by extrapolating the light-dressed transition wavenumber to the limit of zero dressing light intensity. As expected and seen in Fig. 4.4, linear extrapolation might be pursued if data points at low dressing-light intensities are used. On the other hand, by increasing the dressing-light intensity above a certain level, the relation between intensity and tran-

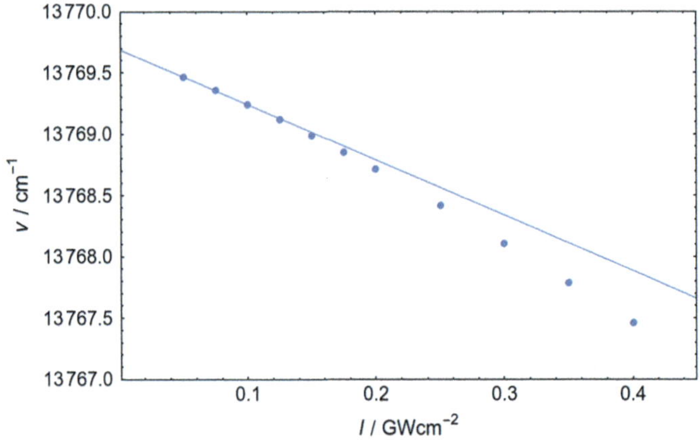

Fig. 4.4 Position of the stimulated emission peak near 13,770 cm^{-1} as a function of dressing light intensity for a $\lambda = 662$ nm laser light. The straight line plotted was obtained by a linear fit to the first four data points

sition wavenumber becomes nonlinear. As seen in Fig. 4.4, the value of the transition wavenumber extrapolated to zero intensity is 13,769.7 cm^{-1}. Considering that the $|X\,0\,0\rangle$ and $|X\,9\,2\rangle$ states involved in this transition belong to the n and $n-1$ Fourier manifolds (see Fig. 4.1), the transition wavenumber between $|X00\rangle$ and $|X92\rangle$ can be calculated to be 1336.0 cm^{-1} = (15,105.7 – 13,769.7) cm^{-1}, where 15,105.7 cm^{-1} is the photon energy of the dressing light. The numerical value for the transition wavenumber, obtained as the difference between the computed field-free eigenenergies, is 1336.1 cm^{-1} = (1415.4 – 79.3) cm^{-1}, where 1415.4 and 79.3 cm^{-1} are the computed energies of the $|X\,9\,2\rangle$ and $|X00\rangle$ states, respectively. Thus, this extrapolation technique works well.

4.4.3 Frequency Dependence of the Light-Dressed Spectrum

Figure 4.5 shows the light-dressed absorption and stimulated emission spectrum of Na_2 dressed by a light field whose intensity is $I = 10^8$ W cm^{-2} at different wavelengths. As can be seen in Fig. 4.5, both the absorption and the stimulated emission spectra vary largely depending on the wavelength of the dressing light. This is the expected behavior because the contribution of different field-free states in the light-dressed states also vary depending on the dressing-light wavelength, leading to varying transition probabilities. Therefore, by changing the dressing-light wavelength, one can influence which transitions appear in the light-dressed spectrum.

Interestingly, the magnitudes of the Franck–Condon overlaps between the vibrational states of the X and A electronic states of Na_2 are reflected in the stimulated emission spectrum, i.e., the number of vertical nodes in the stimulated emission spec-

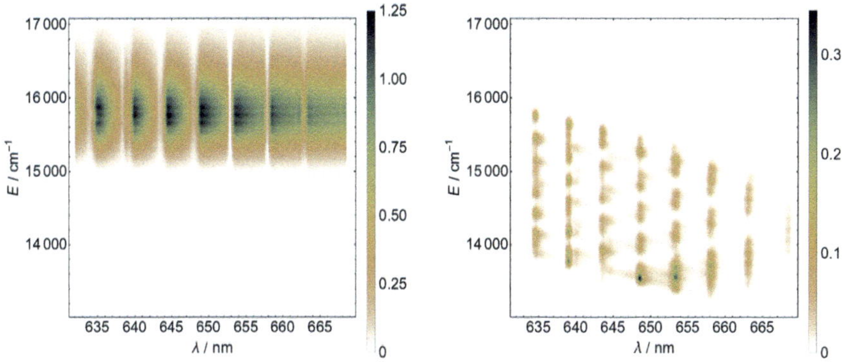

Fig. 4.5 Absorption (left panel) and stimulated emission (right panel) 0 K spectra of Na_2 dressed with an $I = 10^8$ W cm^{-2} intensity laser light of different wavelengths

trum at the different dressing-light wavelengths can reveal which v value of the $|\text{A}\ v\ J\rangle$-type states contributes most to the initial light-dressed state. For example, using dressing light at 662 nm leads to emission lines whose transition amplitude primarily originate from $|\text{A}\ 1\ J\rangle \rightarrow |\text{X}\ v\ J \pm 1\rangle$-type transitions, while using a dressing light with 657 nm leads to emission lines whose transition amplitudes primarily originate from $|\text{A}\ 2\ J\rangle \rightarrow |\text{X}\ v\ J \pm 1\rangle$-type transitions.

4.4.4 *Light-Dressed Spectra at Finite Temperatures*

In the preceding sections it was assumed that the initial light-dressed state correlates to the field-free rovibronic ground state of Na_2, that is, the light-dressed spectra shown are those at $T = 0$ K. The physical picture behind this assumption is that initially the field-free molecules are all in their ground state and these are transformed into dressed states with the adiabatic turn-on of the dressing field. In a realistic experiment at a finite temperature, however, the molecules are not necessarily in their ground state, and thermal averaging of the computed spectrum needs to be carried out. Because thermal equilibrium is assumed prior to the light-dressing process, the thermal averaging can be done by weighting transitions with the Boltzmann weights of the field-free states correlating to the respective initial light-dressed states. That is, transitions from each $|\Phi_i\rangle$ light-dressed state are included in the computed spectrum, and all transitions from a given $|\Phi_i\rangle$ light-dressed state are weighted by

$$\frac{e^{-E_i^{\text{FF}}/kT}}{Q(T)}, \tag{4.41}$$

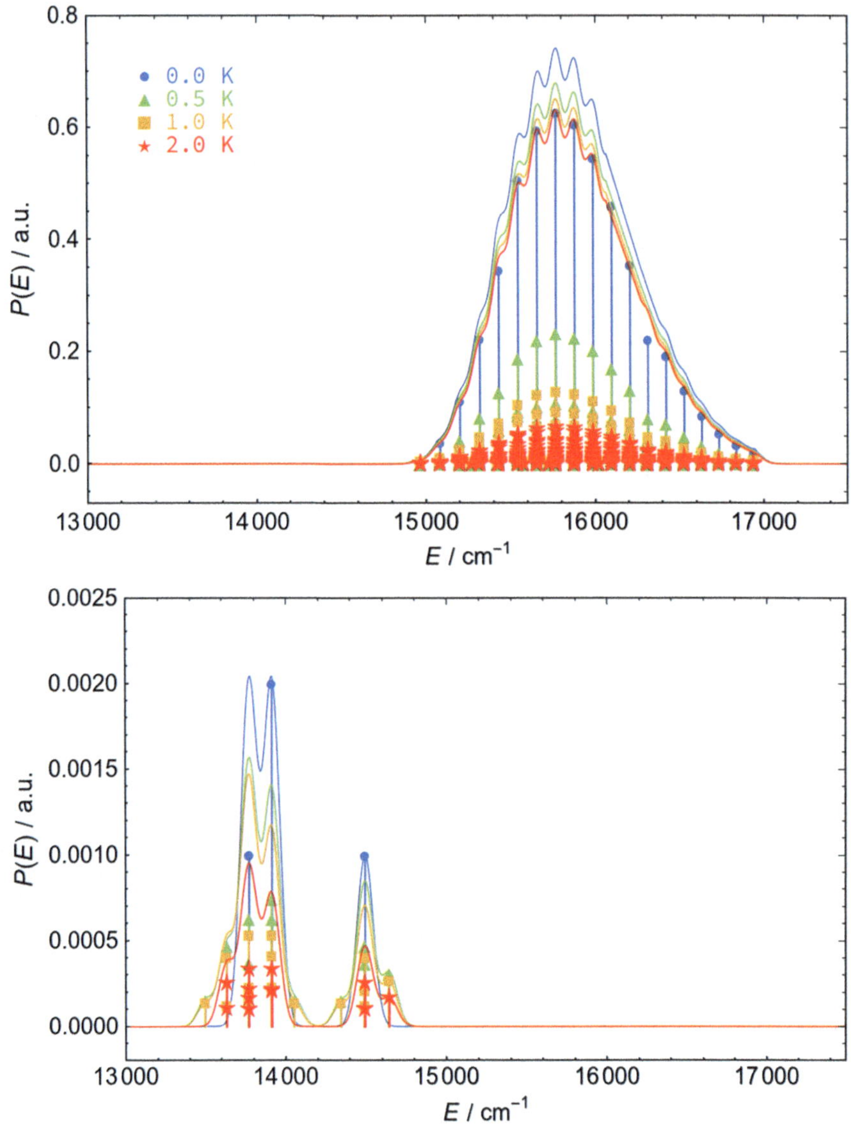

Fig. 4.6 Light-dressed absorption (upper panel) and stimulated emission (lower panel) spectra of Na_2 obtained at different temperatures with a dressing light whose wavelength and intensity are 662 nm and 5×10^7 Wcm^{-2}, respectively

where $Q(T)=\sum_i e^{-E_i^{\mathrm{FF}}/kT}$ is the rovibronic partition function of the field-free molecule, and E_i^{FF} is the energy of the field-free rovibronic state to which $|\Phi_i\rangle$ correlates in the limit of the dressing light intensity going to zero.

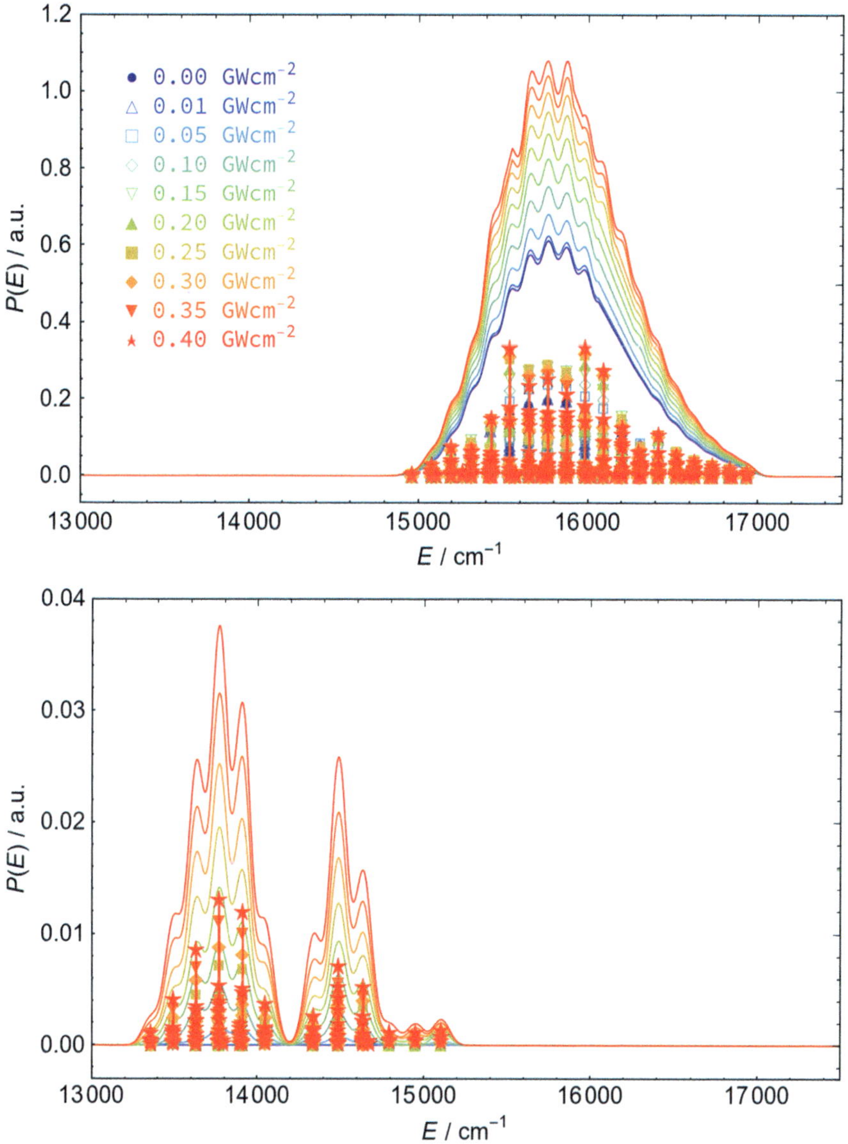

Fig. 4.7 Light-dressed absorption (upper panel) and stimulated emission (lower panel) spectra of Na_2 obtained at 0.5 K with 662 nm wavelength dressing fields having different intensities

At finite temperatures an additional complication arises. Namely, one needs to take into account the field-free rotational states of a closed-shell diatomic molecule which are characterized not only by J but also by the m quantum number, which stands for the projection of the rotational angular momentum onto the chosen space-fixed quantization axis. Since a linearly polarized dressing field (and the weak probe pulse with identical polarization) can not mix states with different m quantum numbers, one can restrict the simulation to the $m = 0$ manifold when $T = 0$ K. In the finite temperature calculations, one needs to determine the light-dressed states and corresponding transitions for the different m manifolds, and include them in the spectrum with appropriate weights given by (4.41).

Figure 4.6 shows the light-dressed spectra of Na_2 at different temperatures when Na_2 is dressed by a light field whose wavelength and intensity are 662 nm and 5×10^7 Wcm^{-2}, respectively. The upper panel of Fig. 4.6 demonstrates that the absorption peaks exhibit splittings and that the height of the split peaks decrease significantly when the temperature increases. This is because the low-lying rotational states are populated at finite temperatures. Nonetheless, the height of the spectrum envelope is much less sensitive to the temperature than the individual peak heights. The effects of a temperature increase appear more significantly in the stimulated emission spectrum than in the absorption spectrum, as shown in the lower panel of Fig. 4.6.

Figure 4.7 shows the dependence of the light-dressed spectra on the dressing light intensity at $T = 0.5$ K. In a similar manner as in Fig. 4.6, the intensities of the spectral peaks in Fig. 4.7 are more influenced by the increase of temperature than the entire spectral envelope.

4.4.5 *Effects of the Dressing-Field Turn-On Time on the Light-Dressed States*

Up to this point it was assumed that the dressing field is turned on adiabatically. As explained in Sect. 4.2.2, the result is that an initial field-free eigenstate is transformed into a single light-dressed state during the dressing process. If the dressing field is not turned on adiabatically, the generated light-dressed wave function becomes a superposition of light-dressed states, with the coefficients depending on the turn-on time [35, 38, 43, 44].

Figure 4.8 demonstrates the population of the different field-free eigenstates in the wave function for dressing-fields of different turn-on time (Γ) and intensity. The functional form of the dressing light was assumed to be $\boldsymbol{E}_1(t) = \boldsymbol{0}$ for $t < 0$, $\boldsymbol{E}_1(t) = \boldsymbol{E}_{\max}\sin(\hbar\omega_1 t)\sin^2(\pi t/\Gamma)$ for $0 < t < \Gamma/2$, and $\boldsymbol{E}_1(t) = \boldsymbol{E}_{\max}$ for $\Gamma/2 < t$. The populations shown in Fig. 4.8 were computed for $t = \Gamma/2$ by solving the TDSE directly.

Based on the $|\mathrm{X}\ 0\ 0\rangle \leftrightarrow |\mathrm{X}\ 0\ 1\rangle$ transition of Na_2, the characteristic timescale of molecular rotations is around 100 ps. Panels (b) and (c) of Fig. 4.8 demonstrate that when the turn-on time is shorter than this characteristic timescale, the degree of rotational excitation is reduced. In fact, in the excited electronic state the rotational excitation is limited to that required by the optical selection rules ($\Delta J = \pm 1$). Panels (b) and (c) of Fig. 4.8 also demonstrate that as the turn-on time of the dressing light

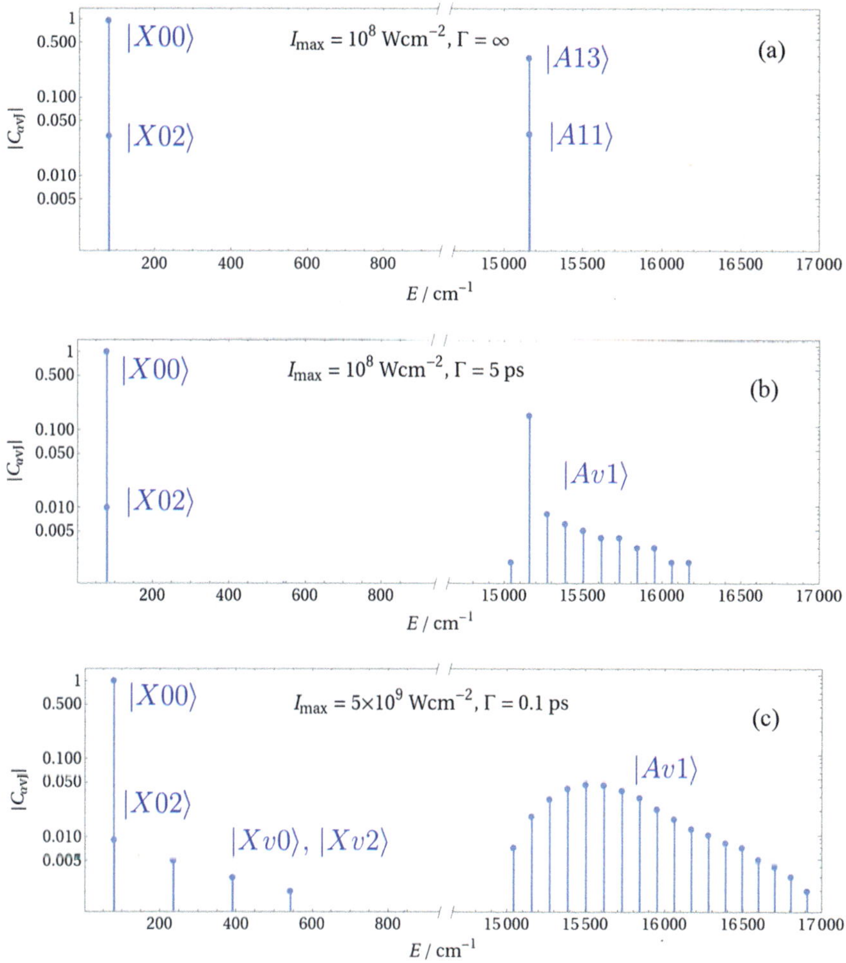

Fig. 4.8 Population of the different field-free eigenstates in the wave function for $\lambda = 663$ nm wavelength dressing-fields of different turn-on time and intensity

becomes shorter, its spectral width expands, giving rise to significant populations in $|A\ v\ 1\rangle$-type field-free eigenstates in a wider energy range.

After the dressing field reaches its peak intensity at $t = \Gamma/2$ the wave function can be expanded as a superposition of the light-dressed states and the light-dressed spectrum can be calculated according to (4.35). If the probe pulse is long enough so that the interferences in the transition probability are averaged out, the light-dressed spectrum can be generated as a simple weighted sum of the spectra of individual light-dressed states.

In summary, by changing the turn-on time of the dressing field, the initial wave function can be manipulated, and in such a way the transition peaks observed in the light-dressed spectrum can be influenced or controlled. However, this might significantly complicate the interpretation of the spectrum.

4.5 Summary and Conclusions

We have presented a theoretical method to compute light-dressed spectra of molecules dressed by medium-intensity laser fields ($I = 10^7 - 5 \times 10^9$ Wcm^{-2} in the numerical examples of this work). The approach is based on Floquet theory for deriving the light-dressed states, which are expanded using field-free molecular eigenstates. Once the light-dressed states are determined, the transition amplitudes induced between them by a weak probe pulse are computed utilizing first-order perturbation theory.

A numerical example is given for the homonuclear diatomic molecule Na_2, for which the general formulae can be simplified and the rovibronic light-dressed spectra can be interpreted straigthforwardly. The simulations indicate that the light dressing process leads to the splitting of field-free peaks as well as the appearance of new peaks, which correspond to transitions forbidden in the field-free case. Processes corresponding to the well-known *intensity borrowing* and *Autler–Townes* effects of molecular spectroscopy can also be observed.

Dependence of the light-dressed rovibronic spectrum of Na_2 on the dressing light intensity and the dressing light wavelength was investigated, as well. It was found that one can influence and control the peaks appearing in the light-dressed spectra by manipulating the intensity and the wavelength of the dressing light.

We tested how the transition wavenumber of a forbidden field-free transition could be determined by extrapolating light-dressed transition frequencies to the limit of zero dressing-light intensity. The extrapolation scheme seemed to produce excellent results.

Calculations assuming different initial temperatures of the molecular ensemble prior to the dressing process reveal that the envelopes of the field-dressed spectra are much less sensitive to the initial temperature than the individual spectral peaks, similar to the field-free case.

Finally, it was shown that the initial wave function of the light-dressed system can be modified significantly when the turn-on time of the dressing field is changed from the adiabatic limit to shorter times, likely to result in significant changes in the light-dressed spectrum, as well.

Acknowledgements This research was supported by the EU-funded Hungarian grant EFOP-3.6.2-16-2017-00005. The authors are grateful to NKFIH for support (Grant No. PD124623 and K119658).

References

1. F. Merkt, M. Quack, Molecular quantum mechanics and molecular spectra, molecular symmetry, and interaction of matter with radiation, in *Handbook of High-Resolution Spectroscopy*, ed. by M. Quack, F. Merkt (Wiley, Chichester, 2011)
2. T. Udem, R. Holzwarth, T.W. Hänsch, Optical frequency metrology. Nature **416**, 233–237 (2002)
3. S.A. Diddams, The evolving optical frequency comb. J. Opt. Soc. Am. B **27**, B51 (2010)
4. J.L. Hall, Nobel lecture: defining and measuring optical frequencies. Rev. Mod. Phys. **78**, 1279–1295 (2006)
5. T.W. Hänsch, Nobel lecture: passion for precision. Rev. Mod. Phys. **78**, 1297–1309 (2006)
6. T. Brabec, F. Krausz, Intense few-cycle laser fields: frontiers of nonlinear optics. Rev. Mod. Phys. **72**, 545–591 (2000)
7. F. Krausz, M. Ivanov, Attosecond physics. Rev. Mod. Phys. **81**, 163–234 (2009)
8. A. Palacios, F. Martín, The quantum chemistry of attosecond molecular science. WIREs Comp. Mol. Sci. 20, e1430 (2020)
9. A.H. Zewail, Femtochemistry: atomic-scale dynamics of the chemical bond. J. Phys. Chem. A **104**, 5660–5694 (2000)
10. M. Dantus, Coherent nonlinear spectroscopy: from femtosecond dynamics to control. Annu. Rev. Phys. Chem. **52**, 639–679 (2001)
11. S. Mukamel, Multidimensional femtosecond correlation spectroscopies of electronic and vibrational excitations. Annu. Rev. Phys. Chem. **51**, 691–729 (2000)
12. T. Ando, A. Iwasaki, K. Yamanouchi, Strong-field fourier transform vibrational spectroscopy of D_2^+ using few-cycle near-infrared laser pulses. Phys. Rev. Lett. **120**, 263002 (2018)
13. C. Cohen-Tannoudji, J. Dupont-Roc, G. Grynberg, *Atom-Photon Interactions: Basic Processes and Applications* (Wiley-VCH Verlag GmbH and Co, KGaA, 2004)
14. J.H. Shirley, Solution of the Schrödinger equation with a Hamiltonian periodic in time. Phys. Rev. **138**, B979–B987 (1965)
15. C. Wunderlich, E. Kobler, H. Figger, T.W. Hänsch, Light-induced molecular potentials. Phys. Rev. Lett. **78**, 2333–2336 (1997)
16. B.M. Garraway, K.-A. Suominen, Adiabatic passage by light-induced potentials in molecules. Phys. Rev. Lett. **80**, 932–935 (1998)
17. Y. Sato, H. Kono, S. Koseki, Y. Fujimura, Description of molecular dynamics in intense laser fields by the time-dependent adiabatic state approach: application to simultaneous two-bond dissociation of CO_2 and its control. J. Am. Chem. Soc. **125**(26), 8019–8031 (2003)
18. M.E. Corrales, J. González-Vázquez, G. Balerdi, I.R. Solá, R. de Nalda, L. Bañares, Control of ultrafast molecular photodissociation by laser-field-induced potentials. Nat. Chem. **6**, 785–790 (2014)
19. F. Kelkensberg, C. Lefebvre, W. Siu, O. Ghafur, T.T. Nguyen-Dang, O. Atabek, A. Keller, V. Serov, P. Johnsson, M. Swoboda, T. Remetter, A. L'Huillier, S. Zherebtsov, G. Sansone, E. Benedetti, F. Ferrari, M. Nisoli, F. Lépine, M.F. Kling, M.J.J. Vrakking, Molecular dissociative ionization and wave-packet dynamics studied using two-color XUV and IR pump-probe spectroscopy. Phys. Rev. Lett. **103**, 123005 (2009)
20. O. Atabek, R. Lefebvre, T. Nguyen-Dang, Unstable states in laser assisted and controlled molecular processes, in *Unstable States in the Continuous Spectra, Part I: Analysis, Concepts, Methods, and Results*, pp. 51–104, (Elsevier, Amsterdam, 2010)
21. E.E. Aubanel, J.-M. Gauthier, A.D. Bandrauk, Molecular stabilization and angular distribution in photodissociation of H_2^+ in intense laser fields. Phys. Rev. A **48**, 2145–2152 (1993)
22. A.D. Bandrauk, M.L. Sink, Photodissociation in intense laser fields: predissociation analogy. J. Chem. Phys. **74**, 1110–1117 (1981)
23. P.H. Bucksbaum, A. Zavriyev, H.G. Muller, D.W. Schumacher, Softening of the H_2^+ molecular bond in intense laser fields. Phys. Rev. Lett. **64**, 1883–1886 (1990)
24. S.H. Autler, C.H. Townes, Stark effect in rapidly varying fields. Phys. Rev. **100**, 703–722 (1955)

25. A. Sanli, X. Pan, S. Magnier, J. Huennekens, A.M. Lyyra, E.H. Ahmed, Measurement of the Na_2 $5^1\sigma_g^+ \rightarrow A^1\sigma_u^+$ and $6^1\sigma_g^+ \rightarrow A^1\sigma_u^+$ transition dipole moments using optical-optical double resonance and Autler-Townes spectroscopy. J. Chem. Phys. **147**(20), 204301 (2017)
26. E.H. Ahmed, J. Huennekens, T. Kirova, J. Qi, A.M. Lyyra, The Autler-Townes effect in molecules: observations, theory, and applications. Adv. At., Mol., Opt. Phys. **61**, 467–514 (2012)
27. C.Y. Lee, B.H. Pate, Dressed states of molecules and microwave-infrared double-resonance spectroscopic techniques employing an electric quadrupole focusing field. J. Chem. Phys. **107**(24), 10430–10439 (1997)
28. G.J. Halász, Á. Vibók, M. Šindelka, N. Moiseyev, L.S. Cederbaum, Conical intersections induced by light: Berry phase and wavepacket dynamics. J. Phys. B **44**(17), 175102 (2011)
29. G.J. Halász, M. Šindelka, N. Moiseyev, L.S. Cederbaum, Á. Vibók, Light-induced conical intersections: topological phase, wave packet dynamics, and molecular alignment. J. Phys. Chem. A **116**(11), 2636–2643 (2012)
30. T. Szidarovszky, G.J. Halász, A.G. Császár, L.S. Cederbaum, Á. Vibók, Direct signatures of light-induced conical intersections on the field-dressed spectrum of Na_2. J. Phys. Chem. Lett. **9**, 2739–2745 (2018)
31. T. Szidarovszky, G.J. Halász, A.G. Császár, L.S. Cederbaum, Á. Vibók, Conical intersections induced by quantum light: field-dressed spectra from the weak to the ultrastrong coupling regimes. J. Phys. Chem. Lett. **9**, 6215–6223 (2018)
32. T. Szidarovszky, A.G. Császár, G.J. Halász, Á. Vibók, Rovibronic spectra of molecules dressed by light fields. Phys. Rev. A **100** (2019)
33. S.-I. Chu, D.A. Telnov, Beyond the floquet theorem: generalized Floquet formalisms and quasienergy methods for atomic and molecular multiphoton processes in intense laser fields. Phys. Rep. **390**(1), 1–131 (2004)
34. S.-I. Chu, Recent developments in semiclassical Floquet theories for intense-field multiphoton processes, in *Advances in Atomic and Molecular Physics* (Elsevier, Amsterdam, 1985), pp. 197–253
35. S. Guérin, H.R. Jauslin, Control of quantum dynamics by laser pulses: adiabatic Floquet theory. Adv. Chem. Phys. **125**, 147–267 (2003)
36. C. Lefebvre, T.T. Nguyen-Dang, F. Dion, M.J.J. Vrakking, V.N. Serov, O. Atabek, Attosecond pump-probe transition-state spectroscopy of laser-induced molecular dissociative ionization: adiabatic versus nonadiabatic dressed-state dynamics. Phys. Rev. A **88**, 053416 (2013)
37. C. Fábri, R. Marquardt, A.G. Császár, M. Quack, Controlling tunneling in ammonia isotopomers. J. Chem. Phys. **150**, 014102 (2019)
38. A. Leclerc, D. Viennot, G. Jolicard, R. Lefebvre, O. Atabek, Controlling vibrational cooling with zero-width resonances: an adiabatic floquet approach. Phys. Rev. A **94**, 043409 (2016)
39. P.R. Bunker, P. Jensen, *Molecular Symmetry and Spectroscopy* (NRC Research Press, Ottawa, 1998)
40. S. Magnier, P. Millié, O. Dulieu, F. Masnou-Seeuws, Potential curves for the ground and excited states of the Na_2 molecule up to the (3s+5p) dissociation limit: results of two different effective potential calculations. J. Chem. Phys. **98**, 7113–7125 (1993)
41. W.T. Zemke, K.K. Verma, T. Vu, W.C. Stwalley, An investigation of radiative transition probabilities for the $A^1\sigma_u^+$-$X^1\sigma_g^+$ bands of Na_2. J. Mol. Spectrosc. **85**, 150–176 (1981)
42. T. Szidarovszky, A.G. Császár, G. Czakó, On the efficiency of treating singularities in triatomic variational vibrational computations. The vibrational states of H_3^+ up to dissociation. Phys. Chem. Chem. Phys. **12**, 8373–8386 (2010)
43. H. Xu, E. Lötstedt, A. Iwasaki, K. Yamanouchi, Sub-10-fs population inversion in N_2^+ in air lasing through multiple state coupling. Nat. Commun. **6**, 8347 (2015)
44. Y. Zhang, E. Lötstedt, K. Yamanouchi, Population inversion in a strongly driven two-level system at far-off resonance. J. Phys. B **50**, 185603 (2017)

Chapter 5
Coherent Nonlinear Processes in Metal-Semiconductor Hybrid Nanostructures

Parinda Vasa

Abstract Metal nanostructures supporting surface plasmon polaritons (SPPs), confined electromagnetic field modes at the metal-dielectric interface have been shown to posses significant potential for guiding and manipulating light on the nanoscale. The interaction between these localized electromagnetic fields with the emitters in close proximity has led to the emerging and exciting field of active plasmonics. Recent advances in nanotechnology have enabled fabrication of high quality semiconductor and metal nanostructures, in which several aspects of light-matter interactions, particularly those in coherent and nonlinear regimes have been demonstrated and studied extensively. This article presents discussion on experimental demonstration of some of the coherent nonlinear effects in metal-semiconductor hybrid nanostructures. Studies have demonstrated that these hybrid nanostructures may open up exciting possibilities in realizing novel all-optical, ultrafast nanoscopic devices, nanoscale lasers, quantum information processing and, single-molecule sensing applications.

5.1 Introduction

Coherence is a property that describes the phase correlation between physical quantities associated with a single or multiple waves. Since quantum mechanics associates wave properties with matter, it has become a very important concept in quantum physics and energy transfer processes [1–3]. Coherence is at the core of interference phenomena and is essential in forming wavepackets. Though the simplest manifestation of interference is the addition of two waves generating constructive (bright) or destructive (dark) interference fringes, interference between wavepackets in quantum systems give rise to intriguing and complex physical effects. In case of incoherent addition of light waves, the final result is a sum of intensities in which all information related to relative phases is lost [1, 2].

P. Vasa (✉)
Department of Physics, Indian Institute of Technology Bombay, Mumbai 400076, India
e-mail: parinda@iitb.ac.in

K. Yamanouchi and D. Charalambidis (eds.), *Progress in Ultrafast Intense Laser Science XV*, Topics in Applied Physics 136,
https://doi.org/10.1007/978-3-030-47098-2_5

An emitter in an excited state relaxing via spontaneously emitting a photon in free space is an irreversible process that is the most elementary light-matter interaction [1, 2, 4]. Its correct modelling is possible only within quantum mechanical framework by considering coherent and incoherent additions. The model presented by Victor Weisskopf and Eugene Wigner to explain spontaneous emission is based on the interaction between a single emitter and a continuum of vacuum field fluctuations of the environment. The interaction between an quantum emitter and a single electromagnetic field mode is a coherent process, in which the phase relationship is preserved [2, 4, 5]. Hence it is a reversible process and the energy can flow back and forth between the emitter and the field mode without any losses. However, when an emitter (or a quantum mechanical 2-level system) interacts simultaneously with a continuum of quantum-mechanical harmonic oscillator modes, it becomes an incoherent and irreversible process. The phase relationship here is lost due to the time averaging over the continuum of vacuum states characterized by different frequencies and phases without preserving any correlation [1, 2, 4, 5]. The energy once released to the environment consisting of continuum of modes via emission of photon cannot return to the emitter in a coherent fashion. In other words, the energy is lost to the surroundings.

Optical resonators can modify the frequency distribution of vacuum field fluctuations by selectively enhancing those matching the resonance frequency of the resonator. Thus, a resonator prevents loss of coherence by avoiding simultaneous interaction with the continuum of modes. Therefore, if a two-level system or an emitter characterized by emission frequency ω_{emi} is now placed within a lossless resonator having the identical resonance frequency, $\omega_{\text{res}} = \omega_{\text{emi}}$, the rate of spontaneous emission can be drastically altered. It can be suppressed, enhanced and most interestingly can even be made a reversible process [4–9]. Thus, optical resonators provide an opportunity to investigate several coherent quantum mechanical effects, which otherwise are masked due to the presence of continuum of modes in free space. In more advanced quantum optical model, known as the Jaynes-Cummings model [6, 8–10], the coherent interaction between the emitter and the resonator is described as a two-level system interacting with a quantum harmonic oscillator. Here, the time dependent probability of the emitter being either in the excited (e) state or in the ground (g) state is given by $P_e(t) = \cos^2[\Omega_R\sqrt{(n+1)}t]$ and $P_g(t) = \sin^2[\Omega_R\sqrt{(n+1)}t]$, where n is the number of photons in a resonator mode. If there are no photons present, i.e. in absence of the applied field the probability dynamics is governed by the vacuum field fluctuations associated with the resonator, having maximum probability at ω_{res}. In this case, the oscillatory dynamics of the probability (as in case of two coupled oscillators) occurs at the frequency $2\Omega_R$, where $\Omega_R = \mu\mathcal{E}$ is known as the vacuum Rabi frequency. Here, μ represents the dipole moment of the emitter and $\mathcal{E}$ is the strength of the vacuum field [6, 8–10]. Thus, the optical response of the emitter is drastically different from that in the free space. Most importantly, there is a possibility to tailor it by placing it within a resonator. Another intriguing result is the discrete photon number dependence of Rabi frequency, which is fundamentally different from the classical framework of light-matter interaction. This model can be generalized for an ensemble of emitters interacting with multiple modes. However, in this case the system behaves as that comprising coupled classical oscillators with a Rabi frequency given by $\Omega_R = \sqrt{N}\sqrt{M}\mu\mathcal{E}$, where N and M correspond to the number of emitters and field modes, respectively. Unlike in the case of the interaction with a single mode, Ω_R is independent of the photon number, n present in a mode [6, 8–12].

The fully coherent interaction discussed above is reversible only if the emitter and the resonator are completely isolated from the surroundings or are lossless. Since all emitters are characterized by finite lifetime of the excited state and all resonators have a finite photon lifetime governed by their Q-factor, the duration over which the coherent exchange of energy is observed is limited. Beyond this duration, the energy is released to the surrounding irreversibly. Hence, resonators having longer photon lifetime or high Q-factors are required to observe the coherent light-matter interaction. Depending on the relative values of Rabi frequency and the relaxation rates of the emitter and the resonator, two coupling regimes are defined. If Ω_R is lower compared to either of the relaxation rate, it is classified to be weak coupling regime [5, 6, 8, 9, 13, 14]. If Ω_R is higher or comparable to the relaxation rate, in other words if the coupling energy is higher or comparable to the linewidths of the system, it is said to be strong coupling regime [4–6, 8–10]. Only in strong coupling regime, the coherence is preserved long enough to be able to observe the reversible and periodic transfer of energy between the resonator and the emitter. Apart from the energy relaxation processes, other dephasing and line broadening processes also contribute to the effective linewidths of the emitter and the resonator. To attain strong coupling regime, Ω_R needs to be comparable or larger compared to the total linewidth. To a good approximation, the condition for strong coupling is given as $\Omega_R^2 > \frac{\gamma_{\mathrm{emi}}^2}{2} + \frac{\Gamma_{\mathrm{res}}^2}{2}$. Here, γ_{emi} (Γ_{res}) represent the effective linewdith of the emitter (resonator) [8, 9, 14]. Various types of hybrid structures exhibiting weak as well as strong coupling have been widely investigated. Since only in the strong coupling regime, the spectral response is distinctly different, it provides opportunity to investigate coherent nonlinear response drastically different from the individual sub-system, we shall focus only on the optical response of strongly coupled systems. In condensed matter systems, the strong coupling was observed for the first time by Weisbuch et al. in 1992 in inorganic semiconductor micro-cavity at cryogenic temperatures [15]. Later, the coupled mode formation was observed with much higher Rabi energy in organic molecules, which made it possible to explore the phenomena at room temperature. Since then it has been studied in various types of organic systems [8, 9, 14–27]. Those comprising J-aggregated organic molecules are of particular interest because they exhibit very large dipole moment [28]. The origin of this unusually large dipole moment is the coherent coupling between the self-assembled closely packed dye molecules [8, 9, 14, 17–21, 25–27]. Some other photochromic dye molecules, like spiropyran (SPI) also have unusually large transition dipole moments and can reach exceptionally high normal mode splitting values of $\sim$650 meV [24]. Since the observed a splitting corresponds to $\sim$30% of the molecular transition energy (2.2 eV), it is suggestive of the possibility to reach ultra-strong coupling regime. This is an emerging field in quantum optics and requires a new theoretical framework as some of the approximations made in deriving the probability dynamics discussed above breakdown [8, 9, 24].

As we have seen, a resonator is essential to observe coherent light-matter interactions involving vacuum field [5, 6]. All of the light-matter interactions discussed in this Chapter are observed in optical resonators having characteristic length scales comparable to or shorter than the light wavelength. Reflection and interference of light play an important role in designing these resonators. As such a pair of reflecting mirrors is sufficient to construct an optical resonator or a Fabry-Perot cavity, which is the simplest resonator. However, the performance or the Q-factor is significantly improved when instead of a single pair of mirrors, multi-layers of several pairs of alternating transparent (dielectric) materials with different refractive indices are used as in Bragg reflector. Using modern fabrication technology Q-factors as high as 10^6 have been demonstrated for Bragg reflectors [6–9, 14]. Designing and fabricating optical resonators involve ability to manipulate optical fields and light propagation, which is central to optics and photonics. Traditionally, the flow of electromagnetic energy is controlled using elements such as mirrors, lenses, fibres and diffractive elements. These components are diffraction limited posing a significant challenge to highest attainable spatial resolution, miniaturization and high-density integration of optical devices. Recently, more complex structures like photonic crystals and metamaterials have been developed to achieve optimal control with sub-wavelength precision. Photonic crystal is a type of Bragg resonator, in which multiple interferences from periodic interfaces confine light in one-, two as well as three-dimensions. Nevertheless, these resonators at best can be on the scale of the wavelength. It is highly desirable to achieve manipulation of optical fields and control of light-mater interactions with ~ 10 nm or even higher precision, as it holds promise for enhancing the performance and efficiency of several photonic functionalities like photo-detection, -emission and chemical sensing [6, 7]. The main approach to circumvent diffraction limit over a broad range of wavelengths is to exploit surface plasmon polaritons (SPPs) in sub-wavelength metal structures. SPPs are hybrid modes of light waves coupled to free electron oscillations in a metal that can be laterally confined below the diffraction limit [29–32]. The spatial extent of the SPP fields is dictated by the geometry of the metallic nanostructures rather than by the wavelength of the light. The unique optical properties of metals allow field localization on the scale of the skin depth of the metal, which remains sub-micron throughout the broad spectral range. By choosing appropriate geometries, e.g. a gap waveguide, metallic nanowires or chains of metallic nanoparticles as novel SPP waveguides that can transport light on the nanoscale, the ultimate achievable localization is typically below ~ 1 nm, limited by the charge-screening length or the Thomas-Fermi screening length of metals [8, 9, 14, 29–32]. Apart from possible manipulation of light, focusing with ~ 10 nm or even higher resolution could result in a strong field enhancement, enabling efficient manipulation of light-matter interaction and boost optical nonlinearities. For example, plasmonic substrates are vital for the enormous signal enhancement achieved in surface enhanced Raman spectroscopy (SERS)-a technique that can detect a single molecule [33, 34], for bio-sensing and in magneto-optical effects like inverse Faraday effect [8, 14, 30].

Even though the optical properties of SPPs have been studied for many years, there has been an upsurge of new interest in active plasmonics, dealing with metal-semiconductor hybrid structures during the last decade. Particularly, light-matter interaction between SPPs and quantum emitters like excitons in semiconductors has become an active topic of research [8, 14, 30]. Plasmonic fields can be tailored with sub-optical wavelength resolution offering promising scenarios for trapping, guiding and manipulating quantum emitters in dipole potentials, even potentially enabling strong coupling between a single emitter and a single SPP. Such hybrid systems connect two fundamentally different quantum worlds of electromagnetic fields and emitters and open up new opportunities for exploring intriguing light-matter interactions like Purcell effect, vacuum Rabi splitting and the optical Stark effect on nanoscale. Recently, coupling strengths of as high as 350 meV have been observed in metal-semiconductor hybrid nanostructures [8, 14, 24]. Therefore, it is possible to observe light-matter interaction in these hybrid nanostructures even at room temperature without any need for a closed cavity. One of the central goals for hybrid systems is to control light via light by efficiently generating non-classical entangled states of light and matter. Fundamentally new types of all-optical devices, e.g. quantum gates and single-photon transistors have been already envisaged based on coherent nonlinearity in these hybrid structures. Another related unexplored field is the ability of strong coupling to modify the electromagnetic environment of an emitter to modify potential energy landscapes and to control chemical reactions. SPPs can also extend energy transfer range between two emitters, which otherwise is only a few nm, To-date there appears to be no reports of how this resonant energy transfer is modified in presence of coherent effects like strong coupling [8, 9, 14].

Here, we discuss the linear and nonlinear optical response of a metal semiconductor hybrid nanostructures comprising a layer of J-aggregated dye molecules in close proximity of a nano-slit array in a gold film schematically shown in Fig. 5.1a. The hybrid nanostructures consist of a $\sim$50 nm thick J-aggregated cyanine dye (2,2'dimethyl-8-phenyl-5,6,5',6'-dibenzothiacarbocyanine chloride, Hayashibara Biochemicals Laboratories, Inc.) layer spin-coated onto a periodic nano-groove array with 380–460 nm period in a gold film [8, 9, 14, 19, 20, 23, 25–27]. This system exhibits strong exciton-SPP coupling at room temperature hence is a potential candidate to explore coherent nonlinear response in such hybrid structures. We start by reviewing the linear optical response followed by few examples of experiments to investigate coherent nonlinearity. Since the system under consideration consists of an ensemble of molecules interacting with multiple SPP bands, the system behaves as a collection of classical oscillators. Hence the photon number (intensity) dependent Rabi frequency discussed earlier is not expected [8, 9, 14]. Nevertheless, as we shall see in the following sections, the system offers unique possibility to control the Rabi frequency as a function of pump intensity. Similar studies on how quantum emitters and SPPs interact in hybrid structures may lead to advances in several directions and will certainly unravel intriguing phenomena.

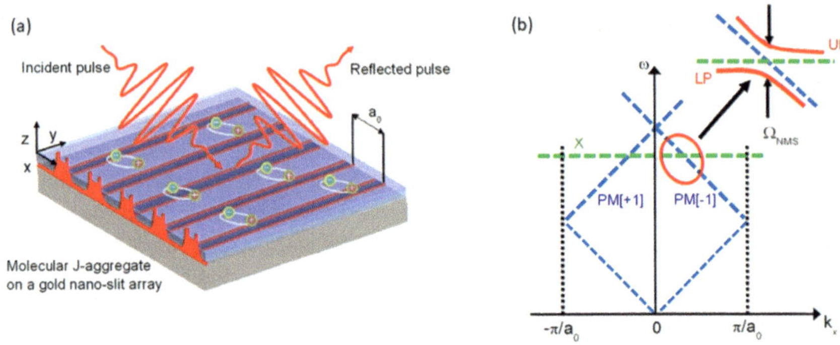

Fig. 5.1 **a** Schematic of a metal-semiconductor hybrid nanostructure used for investigating linear and nonlinear response. It consists of a a gold nano-slit array having period in the range of 380–460 nm, coated with a ~50 nm thick film of J-aggregate molecules in a polymer matrix. Strongly localized SPP fields exist in and near the slits. The optical response of the sample in linear as well as nonlinear regime is investigated by recording angle dependent reflectivity spectra [25]. **b** Schematic of the dispersion relations of the surface plasmon polariton (SPP) resonances excited on a nano-slit array at the polymer-metal interface (PM[±1]) and J-aggregate exciton. The strong dipolar coupling between exciton and SPP results in the formation of upper (UP) and lower (LP) exciton surface plasmon polaritons with normal mode splitting Ω_{NMS} ~60–110 meV [25]. (**a**, **b**) Copyright 2013 Nature Publishing Group

5.2 Linear Optical Response of Metal-J-Aggregate Hybrid Nanostructures

The optical response of a strongly coupled metal-semiconductor hybrid system can be probed without altering the optical response by external light having amplitude lower than the vacuum field associated with the metal resonator. The optical response reveals two distinct resonances, which are formed because of the radiative interaction between excitons and SPPs [8, 9, 14, 23, 25]. These are the hybrid modes that have partly SPP and partly excitonic character. The mode with higher energy is the upper exciton-SPP polariton (UP), whereas the one with lower energy is the lower polariton (LP). They are separated in frequency by normal mode splitting ($\Omega_{\mathrm{NMS}} = 2\Omega_R$) as depicted in Fig. 5.1b. As we shall see in the following sections, the coupled exciton-SPP mode formation is also accompanied by substantial changes in dynamic and nonlinear response.

As the hybrid modes are formed due to the mixing of exciton and SPP modes, it is expected that at resonance condition ($\omega_{\mathrm{exc}} = \omega_{\mathrm{SPP}}$), the modes have equal contribution of exciton and SPP modes. Accordingly, under this condition both LP and UP will have relaxation rate that is the average of individual exciton and SPP relaxation rates [8, 9, 14, 23, 25, 26]. In a coupled system there is yet another possibility that can affect the damping of the mixed modes [35, 36]. While considering the periodic exchange earlier, only coherent exchange was considered in which there is no loss of phase information. Nonetheless, it is possible that either the exciton or the SPP oscillator emits a photon to surrounding via spontaneous emission (loss of phase information), which is absorbed by the other oscillator. This incoherent exchange affects the damping of the coupled modes by an amount that is given by $\sqrt{\Gamma_{\text{exc}}\Gamma_{\text{SPP}}}$ [35, 36]. Such co-operative damping effects have been observed in various systems like trapped ions and molecular aggregates. Though the coupled mode formation in metal-semiconductor hybrid nanostructures has been investigated extensively, the co-operative damping and the dynamics of the coupled modes has been challenging.

Before discussing the experimental results, we briefly discuss the framework of a phenomenological coupled oscillator model to describe the optical response of the metal-J-aggregate hybrid nanostructures [8, 9, 14, 22, 23, 25, 26]. It is a straightforward method similar to that used to describe coupled mechanical harmonic oscillators, yet it can account for co-operative damping phenomena and it is also possible to add some of the other quantum mechanical effects. It is effective to rationalize the observations of linear optical response but needs to be extended to more advanced quantum mechanical Bloch equation formalism for applying to nonlinear response [8, 25–27]. Here, we assume that both emitter and resonator are two-level systems, i.e. each one can either be in the ground state or the excited state. The states corresponding to both of them in the ground state or the excited state are forbidden. Due to the presence of damping, the energy finally leaks out of the system. Thus, the non-hermitian Hamiltonian representing the dipole light-matter interaction can in the form of a 2 × 2 matrix as

$$\widehat{H}_{\text{exc-SPP}} = \hbar\left[\begin{pmatrix} \omega_{\text{exc}} & \Omega_R \\ \Omega_R^* & \omega_{\text{SPP}} \end{pmatrix} - i\begin{pmatrix} \Gamma'_{\text{exc}} & \Gamma_R \\ \Gamma_R & \Gamma_{\text{SPP}} \end{pmatrix}\right]. \tag{5.1}$$

Here, the first matrix of the hamiltonian, $\widehat{H}_{\text{exc-SPP}}$ represents the coherent periodic energy exchange governed by Ω_R, whereas the second matrix represents the incoherent transfer of energy to the surrounding or radiative losses via spontaneous emission [8, 25–27]. The term Γ_R represents the co-operative damping given by $\sqrt{\Gamma_{\text{exc}}\Gamma_{\text{SPP}}}$. Since excitons in an ensemble exhibit considerable non-radiative damping, $\Gamma'_{\text{exc}} = \Gamma_{\text{exc}} + \Gamma^{\text{nr}}_{\text{exc}}$ denotes the sum of the radiative and non-radiative decay rates of the excitons and Γ_{SPP} denotes the radiative damping of the SPP. It has been shown that the major contribution to damping in metal nanostructures is because of radiative damping [37, 38]. Hence, it assumed that there are no non-radiative losses within the resonator (no light is absorbed by the metal nanostructure). The complex eigenfrequencies of the UP and LP modes, $\widetilde{\omega}_{UP,LP}$ are then given by the eigenvalues of $\widehat{H}_{\text{exc-SPP}}$ as $\widetilde{\omega}_{UP,LP} = \frac{1}{2}\left(\omega_{\text{exc}} + \omega_{\text{SPP}} - i(\Gamma'_{\text{exc}} + \Gamma_{\text{SPP}}) \pm A_0\right)$, where $A_0 = \sqrt{\Delta^2 + 4A_1A_2}$ and $\Delta = \omega_{\text{exc}} - \omega_{\text{SPP}} - i(\Gamma'_{\text{exc}} - \Gamma_{\text{SPP}})$. Here $A_1 = \Omega_R - i\Gamma_R$, and $A_2 = \Omega_R^* - i\Gamma_R$. The UP and LP polariton wavefunctions are then obtained as the normalized eigenvectors of $\widehat{H}_{\text{exc-SPP}}$: $|UP, LP\rangle = \frac{(\Delta \pm A_0)|10\rangle + 2A_2|01\rangle}{B_{UP,LP}}$ with a normalization constant $B_{UP,LP} = \sqrt{|\Delta \pm A_0|^2 + 4|A_2|^2}$. Since we are considering the interaction in linear regime, the terms corresponding to both the exciton and the SPP in ground state or in the excited state are omitted. Using this coupled oscillator model in conjunction with the first order perturbation theory, it is possible to evaluate the polariton dipole moments as $\mu_{UP,LP} = \frac{(\Delta \pm A_0)\mu_{\text{exc}} + 2A_2\mu_{\text{SPP}}}{B_{UP,LP}}$, where μ_{exc} and μ_{SPP} are the exciton and SPP dipole moments (oscillator strengths), respectively [8, 9, 25–27]. Though this phenomenological model is simple to implement and is widely used, the major drawback is that the damping is introduced phenomenologically using experimental data. Also, the inhomogeneous broadening has to be accounted by convoluting the optical response with appropriate distribution function.

Modeling of nonlinear optical response and the dynamics of the strongly coupled system is more complex as well as challenging. It at least requires the semi-classical approach based on Liouville equation and density matrix formalism [8, 25–27]. Instead of using phenomenological approach, the damping here can be introduced using Lindblad formalism [2, 35, 39]. As we shall see in the later sections, this semi-classical model can satisfactorily account for the experimental observations (such as Rabi oscillations and optical Stark effect), and make predictions about the phenomena yet to be observed (like higher rungs of the Jaynes-Cummings ladder and electromagnetically induced transparency).

The linear optical response of the hybrid nanostructure is investigated by recording angle resolved *p*-polarized broadband reflectivity (1.6–2 eV) spectra using an ultrafast white-light source (Fianium SC-450-4) [23, 25, 26]. A complete characterization of the emitted in terms of amplitude and phase is also possible using spectral interferometry technique. Fully characterizing the response in terms of amplitude and phase permits better estimation of homogeneous and inhomogeneous broadening. The well-studied linear optical response of such hybrid nanostructures shown in Fig. 5.2a is governed by the strong coupling of excitonic transition dipole moments to vacuum fluctuations of the groove array SPP modes. Due to this coupling, the original J-aggregate and SPP resonances are transformed into strongly coupled exciton-SPP polariton modes (LP and UP), exhibiting a characteristic anti-crossing with a normal mode splitting, $|\Omega_{\text{NMS}}|$ ~60–110 meV [23, 25–27]. The linear

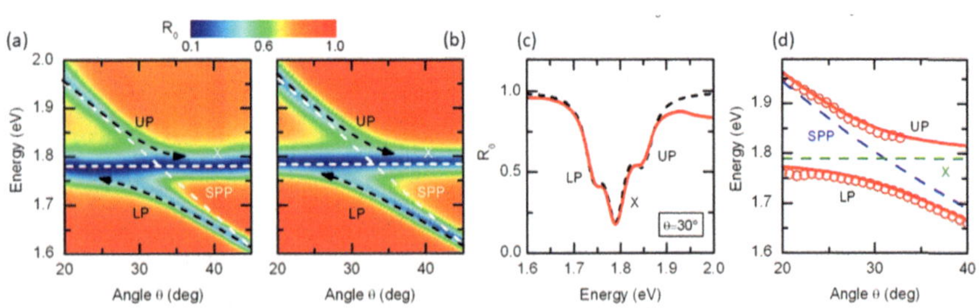

Fig. 5.2 **a** Measured and **b** simulated angle-resolved, *p*-polarized linear reflectivity spectra showing a clear anti-crossing and $\Omega_{\text{NMS}} \sim 110$ meV for a hybrid nanostructure having grating period of 430 nm [25]. The simulations are based on the coupled oscillator model and Bloch equations. The dispersion-less feature at 1.789 eV arises from uncoupled J-aggregate molecules. **c** Comparison between the measured (solid line) and simulated reflectivity spectrum near the anti-crossing [25]. **d** Comparison between the experimentally obtained (open circles) and simulated exciton-SPP polariton dispersion relations [25]. (**a**–**d**) Copyright 2013 Nature Publishing Group

optical properties of the coupled system are well explained within the framework of a coupled oscillator model discussed above. Since in this case, it is an ensemble of molecules interacting with multiple SPP modes, the dipole coupling is given by $\Omega_{NMS} = 2\int \boldsymbol{\mu}_{\mathbf{eff}}(\mathbf{r}) \cdot \mathbf{E}_{\text{SPP}}(\mathbf{r})d^3r$. Here, $\boldsymbol{\mu}_{\mathbf{eff}} \propto \sqrt{N}$ denotes an effective dipole moment density and N is the exciton number. $\mathbf{E}_{\mathbf{SPP}}(\mathbf{r})$ gives the average strength of the local SPP vacuum electric field fluctuations at position $\mathbf{r}$ [23, 25–27]. The simulated reflectivity spectra obtained using the coupled oscillator model and Bloch equations are shown in Fig. 5.2b [25–27]. The experimentally obtained and calculated reflectivity spectra near the anti-crossing and the dispersion relations of the coupled modes are compared in Fig. 5.2c and d, respectively.

At the zero exciton-SPP detuning ($\sim$30°), both polaritons contain 50% exciton and SPP fractions however, the collective polariton modes display strikingly different amplitudes and widths under resonant condition. The difference can arise due to the co-operative damping arising due to the inherent damping of excitons and SPPs [26, 35, 36]. The difference in the damping of the collective modes is known as sub- and super-radiance. Such co-operative damping has been observed in many strongly coupled systems like trapped ions, QDs, and SPPs [8, 26, 35, 36, 40]. The experimental results strongly suggest that the optical properties of the hybrid systems are largely governed by the interplay between the coherent as well as the incoherent energy exchange processes between excitons and SPPs. Even though excitons and SPPs are fundamentally very different, the co-operative damping opens up possibility of tailoring the polariton dynamics. Since the radiative damping of SPPs is much faster compared to excitons, the co-operative damping effects are relatively weaker. If both subsystems are resonant and have equal radiative damping rates ($\Gamma_{\text{exc}} = \Gamma_{\text{SPP}}$), one of the polariton mode will decay with very fast rate, whereas another will not decay at all. In other words, one of the modes is transformed into a perfectly dark mode (with zero decay rate), which is completely decoupled from the environment [8, 26, 35]. Thus, co-operative damping can lead to enhancement or total suppression of spontaneous emission or population trapping induced completely by vacuum field fluctuations. Such a tailoring of polariton dynamics is promising for developing quantum plasmonic devices [41–44].

5.3 Polariton Dynamics

We now turn to nonlinear response of the metal-J-aggregate hybrid nanostructure. As mentioned in the earlier section, in an ensemble of emitters interacting with multiple SPP modes, the Rabi frequency is independent of the number of photons present in each mode. Hence one might think that metal-J-aggregate hybrid structures are not likely to exhibit any nonlinear response. On the other hand, exciton and SPP are fundamentally different with exciton exhibiting a strong optical nonlinearity. Since polaritons are formed due to mixture of excitons and SPPs, polaritons also exhibit nonlinearity like exciton. In this section we discuss experiments performed to investigate incoherent as well coherent polariton dynamics using nonlinear spectroscopy.

To investigate the incoherent dynamics, the differential reflectivity $\Delta R/R_0$ (where, ΔR is the difference in reflectivity in presence and absence of pump pulse and R_0 is the sample reflectivity in absence of the pump pulse) measurements are performed using a pump-probe set-up based on an optical parametric amplifier system operating at a repetition rate of 1 kHz [8, 9, 23]. 80-fs pump pulses centered at 620 nm are used for excitation and a white-light continuum generated in a sapphire plate provides the time-delayed probe pulses covering the range from 630 to 750 nm (schematically depicted in Fig. 5.3a). The pump wavelength is chosen such that it is non-resonant with the J-aggregate as well as the SPP mode of the grating for the chosen range of incidence angles. This leads to incoherent excitation of the J-aggregates via higher energy levels [28]. The optical pumping here mainly results in altering the exciton population via incoherent population transfer and relaxation, effect of which is monitored by recording the differential reflectivity spectra as a function of pump-probe delay. Both pump and probe pulses are *p*-polarized and nearly collinearly focused onto the sample with a beam diameter of $\sim$100 μm at an incidence angle θ with respect to the sample normal. Angle-resolved $\Delta R/R_0$ spectra from 30° to 50° are recorded at different pump-probe delays Δt under vacuum at a temperature of $T = 77$ K to minimize J-aggregate photobleaching. Control experiments are performed on a planar dye-coated gold sample without nanoslit arrays [8, 23].

Angle-resolved $\Delta R/R_0$ spectra recorded at pump-probe delay of 150 fs and pump fluence of 9 μJ/cm^2 are shown in Fig. 5.3b exhibit pronounced nonlinearity near the polariton resonance energies even at angles and wavelength considerably away from the exciton-SPP crossing. Under the experimental conditions used in these experiments, SPP modes essentially behave like a linear oscillator and do not contribute to the nonlinear response [8, 14]. The remarkable observation of polariton nonlinearity clearly demonstrates that the strong radiative interaction between excitons and SPPs induces nonlinearity in the coupled mode even in case of a ensemble. Interestingly, instead of the saturation of absorption of J-aggregate, with enhanced reflectivity at polariton resonance, the polariton nonlinearity ΔR exhibits a dispersive line shape at all pump fluences and incidence angles. As shown in Fig. 5.3b, there is a positive nonlinear signal, (enhanced probe reflectivity), at energies slightly above (UP branch) or slightly below (LP branch) the polariton energies. The enhancement is accompanied by reduction in reflectivity at the low-energy (UP) or high-energy (LP) side of $\omega_{UP,LP}$. Apart from the dispersive nonlinear lineshape, the fluence dependence also exhibits reduction in the normal mode splitting. These results support the conclusion that most of the strong polariton nonlinearities seen in Fig. 5.3a, b result from a saturation of Ω_{NMS} [8, 14, 23, 25–27]. A qualitative lineshape analysis along with the expression for polariton frequencies, suggest that the origin of the nonlinear response is the transient reduction in Ω_{NMS} (Fig. 5.3a).

The observed nonlinear response is attributed to the effect of the incoherent optical pumping on Ω_{NMS}. The pump-generated exciton density saturates the exciton number which, in turn transiently reduces the Ω_{NMS}. Thus, the pump-induced exciton creation results in a time-dependent normal mode splitting $\Omega'_{\text{NMS}}(t) = \Omega_{\text{NMS}}\sqrt{n_0(t) - n_1(t)}$, where Ω_{NMS} is the normal mode splitting observed in the weak excitation limit and

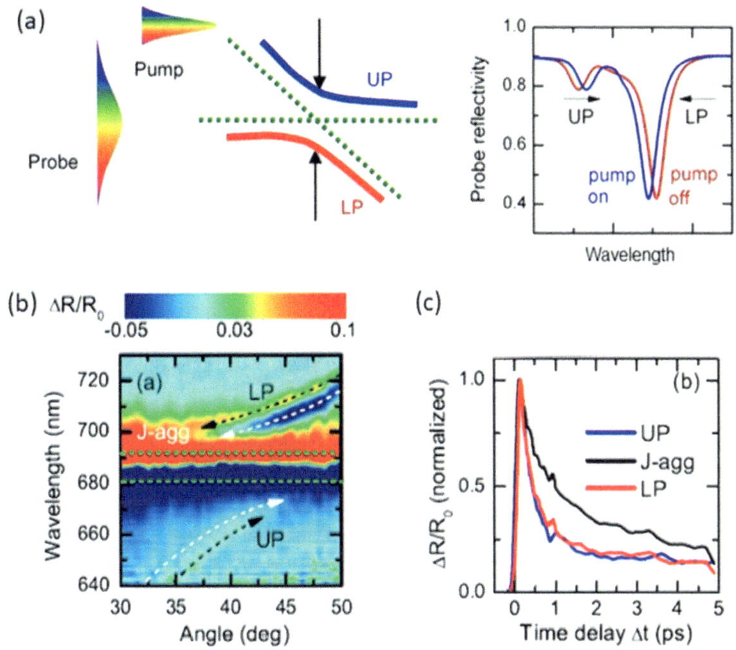

Fig. 5.3 **a** Schematic of a transient reflectivity measurement scheme. A blue-shifted, off-resonant pump pulse transiently shifts the polariton levels. The pump induced reduction in Ω_{NMS} results in significant but incoherent transient changes in probe reflectivity on ultrafast timescale [8]. **b** Angle-resolved $\Delta R/R_0$ spectra ($T = 77$ K) of the hybrid structure with a grating period of 380 nm [23]. The sample is excited by 80 fs pulses centered at 620 nm. The spectra are recorded in a nearly collinear pump-probe set-up at a delay of 150 fs. Results are presented for pump energies of 1.4 nJ, which correspond to fluence of $F_{Pu} = 9$ μJ/cm^2. White dashed lines correspond to angle-dependent non-linear exciton-SPP polariton response, whereas the black dashed lines mark the polariton features in absence of pump (linear regime). An angle-independent response arising from uncoupled J-aggregate molecules (green dashed lines) is also observed. **c** Time dynamics of the normalized $\Delta R/R_0$ signal at the UP (675 nm), LP (710 nm) and exciton (685 nm) resonances corresponding to $F_{Pu} = 159$ μJ/cm^2 [23]. The short UP and LP lifetimes reflect the pronounced radiative polariton damping. (**a**) Copyright 2018 American Chemical Society, (**b**, **c**) Copyright 2010 American Chemical Society

n_0 and n_1 are the exciton populations in the ground-state and excited-state, respectively. A similar transient reduction in dipole coupling has been observed in semiconductor microcavities [12, 20, 45, 46]. As we shall see in the following sections, pump-induced transient reduction in normal mode splitting due to the saturation of exciton number is the most prominent nonlinearity observed in such extended metal-J-aggregate hybrid nanostructures.

Figure 5.3c compares the observed time dynamics of the normalized $\Delta R/R_0$ signal at the bare exciton, and the upper and the lower polariton resonances. It reveals an exciton lifetime of less than 0.9 ps, typical for this kind of J-aggregated dye. The polariton lifetimes are substantially shorter, less than 0.4 ps. The relatively fast relaxation of polariton is expected due to the extremely fast relaxation rate of SPPs [8, 14, 37, 38]. Near the anti-crossing, polaritons are expected to have the average relaxation rate of exciton and SPP. Since exciton relaxation rate is much slower, polariton relaxation should be comparable to that of SPPs. However, the observed relaxation dynamics is similar to that of the uncoupled excitons. The relatively slower polariton dynamics is presumably due to the incoherent excitation conditions used in these experiments [8, 47–50]. There is continuous population transfer occurring from higher energy states to polariton states within 1 ps, slowing down the polariton relaxation. In contrast to the faster polariton relaxation observed in these experiments, Schwartz et al. have reported longer relaxation times for LP mode in J-aggregate/metallic planar cavity hybrid structure under nonresonant excitation [51]. Such slower dynamics can also arise due to motional narrowing [52], and incoherent population transfer from other states or dark mode formation. These results suggest that to get the correct estimate of the polariton dynamics, it is essential to use coherent excitation conditions [47–49]. Nevertheless, the observation of a pronounced, externally-induced, fast and reversible change in reflectivity of a metal-molecular aggregate nanostructure suggests that these hybrid structures can indeed act and potentially find applications as externally controllable, ultrafast switchable mirrors [42, 44].

Recently, Wang et al. have investigated the relation between strong coupling and the radiative damping of the coupled modes in hybrid structures both in the spectral and temporal domains [8, 9, 14, 26]. Under resonant excitation conditions, in which the pump wavelength is resonant with the coupled modes, the LP nonlinear response (Fig. 5.4a) mainly reflects the saturation of the polariton exciton-fraction and a weak transient reduction in the Ω_{NMS}. The dynamics of $\Delta R/R_0$ corresponding to the J-aggregate (exciton), LP and the UP resonances along with the bi-exponential fits are depicted in Fig. 5.4b. The exciton population relaxation times in J-aggregate are $\sim$550 fs [28]. The relatively short-lived component observed near zero-delay corresponds to the fast decay of the coherent scattering signal within the exciton polarization relaxation time. Much shorter polariton response times compared to the J-aggregate reflect the pronounced radiative damping of the SPP-fraction [8, 14, 37, 38]. Here, the coherent scattering and the population relaxation times are indistinguishable. For the LP mode, the rapid exponential decay of ΔR with a lifetime of $\sim$55 fs is observed. A weak signal remaining for $\tau > 200$ fs results from probing a small fraction of uncoupled excitons. An even faster decay of the optical nonlinearity is seen when resonantly exciting the UP mode. Here, the ΔR signal decays to less than 0.1 of its maximum value with a decay time of only $\sim$22 fs [8, 26]. The faintly visible non-exponential decay is again attributed to a weak contribution from uncoupled excitons. These observations provide unambiguous evidence for distinctly different population lifetimes of UP and LP modes at zero detuning and resonant excitation [8, 26, 35, 36, 40]. There is no major contribution from the disorder or the incoherent population transfer discussed earlier. They independently confirm the predictions of the coupled oscillator model and thus establish the formation of sub- and super-radiant polariton states in J-aggregate/metal hybrid nanostructures via co-operative damping. Taken together, these investigations strongly suggest that fluctuating vac-

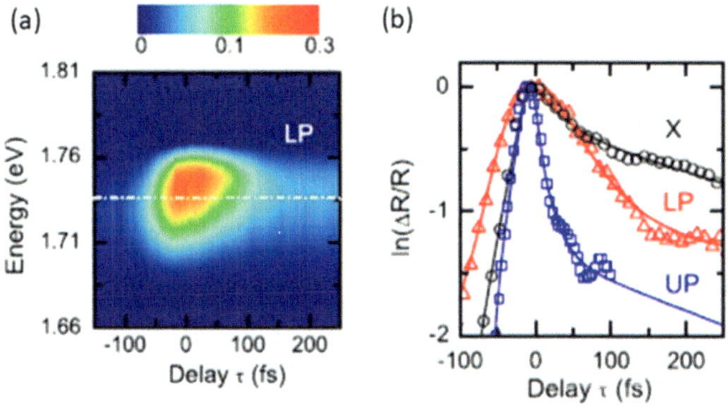

Fig. 5.4 **a** Time evolution of the differential reflectivity signal ($\Delta R/R_0$) for the LP mode of the metal-semiconductor hybrid nanostructure recorded under resonant excitation at $\theta \sim 31°$ [26]. The white line marks the LP resonance energy. Here, the LP nonlinear response mainly reflects the saturation of the exciton fraction. **b** $\Delta R/R_0$ dynamics plotted on the log scale (symbols) for the J-aggregate, the LP (1.74 eV) and the UP (1.84 eV) at zero detuning ($\theta \sim 31°$) [26]. The much shorter UP (22 fs) and LP (55 fs) relaxation times extracted from the bi-exponential fits (lines) compared to the exciton (~550 fs) reflect the strong, radiative damping of the SPPs. Interestingly, the UP relaxes much faster (super-radiant) compared to the LP (sub-radiant), which can occur due to the co-operative damping. (**a**, **b**) Copyright 2014 American Chemical Society

uum fields not only induce a coherent dipole coupling between excitons and SPPs, but also lead to an incoherent exchange of photon energy between both systems [35]. The experimental observations in case of metal-J-aggregate hybrid nanostructures, particularly under resonant excitation are somewhat different from what is known in semiconductor micro-cavities [53, 54]. In these hybrid systems, usually the radiative decay rates of the cavity mode are much longer compared to SPPs and polarization dephasing or inhomogneoues broadening effects are more pronounced. The coherent response of metal-J-aggregate hybrid nanostructures demonstrate that despite the very different nature of both elementary excitations, tailoring of radiative damping phenomena may be an important means to control the polariton dynamics [8, 26].

5.4 Real-Time Observation of Rabi Oscillations

The distinguishing feature of the coherent response associated with metal-semiconductor hybrid nanostructures exhibiting strong coupling is the periodic energy transfer between excitons and SPPs [4–9]. Being ultrafast, it could be promising as an efficient approach to control light on the nannoscale all-optically. Vasa et al. have presented the first real-time observation of ultrafast Rabi oscillations in a molecular aggregate/metal nanostructure [8, 25]. The oscillatory response is again investigated using pump-probe set-up as in earlier experiments. However, in this case both pump and probe pulses have identical spectral profile and are sufficiently broad

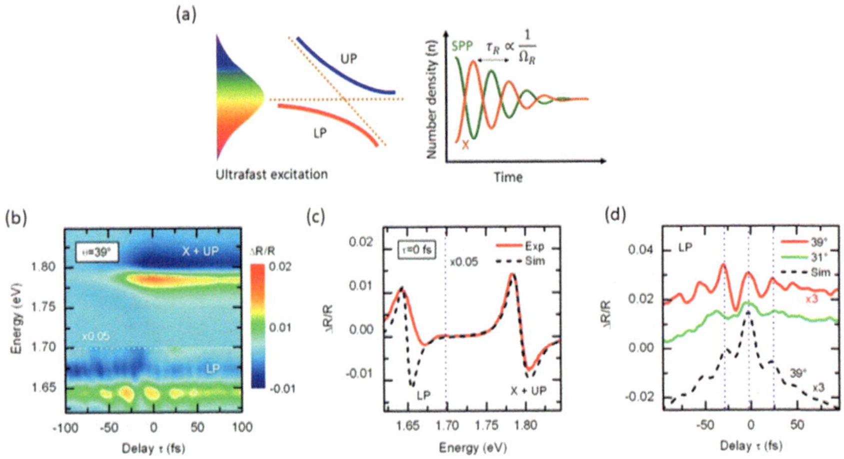

Fig. 5.5 **a** An on-resonance simultaneous excitation of the coupled modes (UP and LP) by a sufficiently short ultrafast, broadband pump pulse initiates a coherent periodic energy transfer between exciton and SPP, resulting in out-of-phase oscillations in their number density [8]. **b** Measured differential reflectivity map for a metal-J-aggregate hybrid nanostructure shown in Fig. 5.1a. It is recorded using nearly-collinearly propagating 15-fs pump and probe pulses. The differential reflectivity ($\Delta R/R_0$) dynamics at the LP resonance clearly reveals a periodic change in SPP population on sub-40 fs timescale corresponding to the real-time observation of Rabi oscillations [25]. **c** Comparison between the measured (solid line) and simulated (dashed line) $\Delta R/R_0$ showing the LP and the uncoupled J-aggregate response [25]. **d** Time evolution of the LP signal corresponding to two different angles exhibiting pronounced Rabi oscillations with period ~30 fs [25]. The shorter oscillation period reflects the larger exciton-SPP detuning. Simulated dynamics (dashed line) for a specific detuning is also shown for comparison. The data are shifted vertically for clarity. **a** Copyright 2018 American Chemical Society, **b–d** Copyright 2013 Nature Publishing Group

to simultaneously overlap with LP as well as UP (Fig. 5.5a). Here too a dispersive spectral lineshape (Fig. 5.5b, c) is observed at the LP [8, 25]. For the given incidence angle, the UP merges with the uncoupled exciton and hence is not distinguishable. The differential reflectivity signal corresponding to LP shows temporal oscillations with a Rabi period of 27 fs both at positive and negative delays (Fig. 5.5d). These coherent oscillations persist for ~200 fs, which is the dephasing time of the polaritons. Since in these samples, SPPs have larger amplitude compared to excitons, the broadband pump pulse launches a coherent polariton wavepacket which, initially, is mainly localized on the SPP state. This wavepacket then coherently oscillates between exciton and SPP states (Fig. 5.5b). The experiments under incoherent excitation discussed earlier show that creation of exciton population in the excited state by pump pulse not only saturates the exciton oscillator but also reduces the normal mode splitting [23]. These two effects combined together explain the observed nonlinear response in Fig. 5.5b–d. The reduction in Ω_{NMS} gives rise to the dispersive lineshape of the LP resonance. Also, it induces larger reduction when the both pump and probe pulses (which are identical in spectral range and nearly co-propagating) create wavepackets localized in the same sub-system. On the other hand, if the two wavepackets are localized in different quantum system, there out-of phase dynamics reduces the nonlinear signal [25]. Thus a stronger differential reflectivity signal is observed at delays corresponding to the integer multiples of the Rabi period. The polariton dynamics discussed here is strongly supported by optical Bloch equations. On increasing the pump pulse energy, the transient suppression in Ω_{NMS} is sufficiently strong to manipulate the Rabi period on ~ 10 fs timescale.

These results present the first real-time observation of coherent exciton-SPP Rabi oscillations in hybrid nanostructures. The oscillations observed in the nonlinear response reflect the modulation in SPP population due to the coherent transfer of energy between SPP excitations of the metallic grating and excitons in the coupled J-aggretate. Remarkably, the dominant nonlinearity governing the nonlinear response under incoherent as well coherent excitation is not the saturation of exciton but the reduction in Ω_{NMS} due to saturation of exciton density. Such a fundamentally different nonlinear response of the hybrid structures can be important for developing novel all-optical ultrafast plasmonic circuits and devices [41–44].

5.5 Optical Starck Effects in Hybrid Systems

The optical Stark effect (OSE) is another fundamental, coherent nonlinear interaction [55–60]. It involves transient shifts of energy levels in the presence of a slightly off-resonant light field. Since it involves coherent response of the system, it is potentially useful for implementing quantum logic and information processing [61, 62]. At least in atomic and some of the semiconductor systems, it can be satisfactorily described by a two-level system interacting with a coherent, transient light field. Recently, OSE has also been explored in metal-semiconductor hybrid nanostructures [56–60]. Along with the transient shift of the resonance, OSE in these hybrid structures is also found to be strongly affected by exciton-SPP interaction [8, 9, 27].

The ultrafast excitation conditions used by Vasa et al. for investigating OSE in metal-Jaggregate hybrid nanostructures [8, 27], is shown in Fig. 5.6a. Under these excitation conditions, three different physical mechanisms are expected to contribute to the transient nonlinearity. First, there may be a saturation of exciton (SPPs do not exhibit significant nonlinearity under these pumping conditions). This saturation should result in saturation of absorption at polariton resonance without affecting its spectral position. However, as in earlier cases, the observed nonlinear response (Fig. 5.6b) is dispersive (transient shift of resonance), These investigations again confirm that saturation of absorption is not the dominant nonlinearity in metal-J-aggregate hybrid structures [8, 23, 26]. Second, the finite exciton population may induce transient reduction in normal mode splitting as discussed earlier. This reduction results in the red-shift of the UP. Also, it results in a red-shift of the UP and thus in a dispersive lineshape with opposite sign of that seen in Fig. 5.6b [8, 23, 25–27]. The third contribution can arise due to the transient Stark-shift of the resonances induced by the off-resonant pump. However, in this case both LP and UP will be blue-shifted, in turn blue-shifting both polariton resonances [8, 27]. The eigen

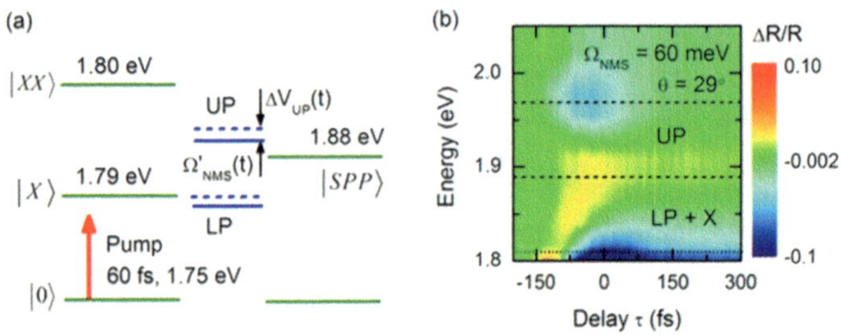

Fig. 5.6 **a** Schematic of the pump-probe experiment performed to explore optical Stark effects in metal-J-aggregate hybrid nanostructure [27], shown in Fig. 5.1a. **b** The differential reflectivity $\Delta R/R_0$ map of the hybrid structure with $\Omega_{\text{NMS}} = 60$ meV recorded near the anti-crossing [27]. For the slightly off-resonant pump pulses centered at 1.75 eV, the map shows a short-lived dispersive signal due to the optical Stark effect at the UP resonance. The dashed lines mark the maxima in $\Delta R/R_0(\tau)$ for the negative and the positive signal. (**a**, **b**) Copyright 2015 American Physical Society

frequencies of the resulting, dressed polaritons can be obtained by diagonalizing an effective Hamiltonian for the exciton-SPP system in presence of an off-resonant driving field. A pure OSE is expected to result in a blue-shift of both resonances, whereas a pure reduction in Ω_{NMS} yields a blue (red) shifted LP (UP). Since, the lineshape in Fig. 5.6b shows the red-shift, for both resonances, the results confirm the OSE in metal-semiconductor hybrid structures [8, 27].

Interestingly, in samples with higher $\Omega_{\text{NMS}} \sim 100$ meV, the results offer an unexpected and very interesting approach for coherently controlling and enhancing OSEs. The strong exciton-SPP interaction in these nanostructures is found to significantly affect the OSEs by changing the direction of the pump-induced shift [62–64]. It has been observed that slightly-off-resonant coherent excitation of the polariton modes, e.g., the LP modes, not only dresses the LP mode, but—by virtue of strong coupling—also transiently Stark shifts the UP mode even though it is greatly off-resonantly (>200 meV) detuned from the pump [8, 27]. In pure OSE, both LP and UP will be shifted in the same directions but here, the shift is caused by the transient reduction in population accompanying OSE and hence the shifts for LP and UP are in opposite directions. These results offer an unexpected and remarkable approach for coherently controlling and enhancing coherent nonlinear response in strongly coupled systems by transiently reducing Ω_{NMS} rather than saturating exciton resonance. Such a strong and coherent nonlinear response is promising for implementing ultrafast coherent, all-optical switching as well as for exploring various aspects of polariton dynamics [8, 14, 41–44].

5.6 Other Coherent Interactions

Since the periodic energy transfer in metal-semiconductor nanostructures occur due to coherent exchange of energy, these structures are promising for exploring coherent phenomena like stimulated emission, lasing and condensation. Apart from periodic energy transfer, Rabi oscillations in a two-level system interacting with light are also interpreted as the system undergoing periodic stimulated emission [4–9]. Since in strong-coupling regime the resonator and the emitter are already interacting coherently, it is expected that these systems will exhibit stimulated emission or lasing at much lower pumping threshold compared to the conventional lasers, which operate in weak coupling regime. The hybrid nature of the polariton mode as contrast to lasing in weak coupling regime has been demonstrated in semiconductor coupled micro-cavities and plasmonic hybrid nanostructures [43, 65, 66]. Some experiments on polariton lasing in metal-semiconductor hybrid nanostructures have recently been reported [67, 68]. Polariton lasing is an example of a collective many-body coherent interaction, which is possible to achieve because of much longer coherence length of polaritons compared to excitons in semiconductors [8, 9, 14, 69]. Another many-body nonlinear interaction that has been observed in strongly coupled microcavities is that of polariton scattering [6]. Like other scattering phenomena such as Raman scattering involving light and lattice vibration, polariton can also interact with other polariton or a different excitation so that they can lose energy as well as momentum via non-radiative mechanism to reach thermodynamically the most favored distribution. Under such a condition, it is possible to have this state highly occupied from which amplified emission can occur. As the excitation and emission are not in the same direction, the emission process here is referred to as polariton scattering. Other nonlinear frequency conversion processes such as parametric generation/amplification and condensation of polaritons involving many-body effects and pulsed operation have also been reported in semiconductor coupled micro-cavities [6]. It would be interesting to explore such coherent interactions in metal-semiconductor hybrid nanostructures. Due to the longer coherence length, exciton-SPP polaritons are also expected to be potential candidates to explore other quantum mechanical many-body interactions like Bose-Einstein condensation that are challenging to be observed otherwise because of their sensitivity to disorder and impurities like in case of purely excitonic systems. Their unique properties make them promising for developing novel types of optical and opto-electronic devices.

5.7 Emerging Trends and Outloook

So far, all coherent as well as incoherent light-matter interactions discussed here have been performed on large ensembles of excitons and with comparatively high pulse energies at least in nanojoule regime. The fundamental quantum mechanical limit in the strong coupling is the dipole interaction between a single exciton and a single SPP mode [8, 9, 70–73]. This regime of vacuum Rabi splitting clearly is much beyond the two strongly coupled classical oscillators or the semi-classical framework for the light-matter interaction. The intrinsically quantum mechanical interaction is expected to offer opportunities to investigate coherent quantum mechanical phenomena on comparatively long length scale and at high temperatures. Two of such phenomena, higher rungs of Jaynes-Cummings ladder and creation of very long lived states isolated from surroundings are briefly mentioned here. Similar effects have been observed in semiconductor coupled micro-cavities and super-conducting systems at very low temperatures [4–9]. The biggest challenges are the strong radiative damping and low coupling energies. Innovative designs of metal nano-antenna and gap modes have been proposed to enhance the field localization and reduce the damping so as to reach the Rabi energies of several meV. The experimental investigations discussed here suggest that considerable progress has been made in understanding the coherent nonlinear response of metal-semiconductor hybrid nanostructures and substantial experimental and theoretical progress seems within the reach with improvements in fabrication techniques, ultrafast spectroscopy, theoretical framework, and experimental as well as theoretical studies. In general, metal-semiconductor hybrid nanostructures provide fascinating possibilities to explore various types of vacuum field induced coherent light-matter interactions at room temperature. Similar investigations were conventionally possible only in very pure atomic systems, high Q-factor microcavities, and superconductors. Due to the plasmonic field enhancement in metallic nanostructures, substantial intensification of light-matter interaction may be possible leading to highly efficient ultrafast coherent switching with femtojoule pulse energies in sub-wavelength structures [41–44]. Ideally, only few or even a single photon would be sufficient to induce the switching in tailor-designed hybrid nanoantennas optimized of strong light-matter interaction. Such ultrafast plasmonic switches and/or transistors would constitute entirely new nano-photonic devices that could create fundamentally different opportunities for all-optical communication and computation [74, 75]. These devices might operate at power levels and switching efficiencies that to date can only be reached with electronic transistors. At the same time, they might operate at extremely high operation speeds in the Terahertz regime, in principle only limited by the femtosecond lifetime of coupled or hybrid mode. Such hybrid exciton/plasmon devices might therefore combine favorable properties of electronic and all-optical devices and would truly merge plasmonics and quantum electronics [6, 8, 9, 14].

References

1. M. Born, E. Wolf, *Principles of Optics*, 7th edn. (Cambridge University Press, Cambridge, 1999)
2. M.O. Scully, M.S. Zubairy, *Quantum Optics* (Cambridge University Press, Cambridge, 2001)
3. P. Vasa, D. Mathur, *Ultrafast Biophotonics* (Springer International Publishing, Switzerland, 2016)
4. G.S. Agarwal, *Quantum optics*, vol. 70 (Springer, New York, 1974)

5. S. Haroche, J.-M. Raimond, *Exploring the Quantum: Atoms, Cavities, and Photons* (Oxford University Press, Oxford, 2006)
6. A.V. Kavokin, J.J. Baumberg, G. Malpuech, F.P. Laussy, *Microcavities* (Oxford University Press, Oxford, 2007)
7. K.J. Vahala, Nature **424**, 839 (2003)
8. P. Vasa, C. Lienau *ACS Photonics* **424**, 2 (2018)
9. P. Vasa, *Encyclepedia of Applied Physics* (Wiely-VCH, in production, 2019)
10. E. Jaynes, F.W. Cummings, Proc. IEEE **51**, 89 (1963)
11. M. Tavis, F.W. Cummings, Phys. Rev. **170**, 379 (1968)
12. G. Khitrova, H.M. Gibbs, F. Jahnke, M. Kira, S.W. Koch, Rev. Mod. Phys. **71**, 1591 (1999)
13. E.M. Purcell, Phys. Rev. **69**, 681 (1946)
14. P. Törmä, W.L. Barnes, Rep. Prog. Phys. **1**, 013901 (2015)
15. C. Weisbuch, M. Nishioka, A. Ishikawa, Y. Arakawa, Phys. Rev. Lett. **69**, 3314 (1992)
16. T. Yoshie, A. Scherer, J. Hendrickson, G. Khitrova, H.M. Gibbs, G. Rupper, C. Ell, O.B. Shchekin, D.G. Deppe, Phys. Rev. Lett. **69**, 3314 (1992)
17. I. Pockrand, A. Brillante, D. Moebius, J. Chem. Phys. **77**, 6289 (1982)
18. J. Bellessa, C. Bonnand, J.C. Plenet, J. Mugnier, Phys. Rev. Lett. **93**, 036404 (2004)
19. J. Dintinger, S. Klein, F. Bustos, W.L. Barnes, T.W. Ebbesen, Phys. Rev. B **71**, 035424 (2005)
20. J. Dintinger, I. Robel, P. Kamat, C. Genet, T. Ebbesen, Adv. Mater. **18**, 1645 (2006)
21. N.T. Fofang, T.H. Park, O. Neumann, N.A. Mirin, P. Nordlander, N.J. Halas, Nano Lett. **8**, 3481 (2008)
22. R. Pomraenke, S. Schwieger, Y.I. Mazur, V. Kunets, P. Srinivasan, E. Johnson, J.E. Kihm, D.S. Kim, E. Runge, G. Salamo, C. Lienau, Phys. Rev. Lett. **101**, 116801 (2008)
23. P. Vasa, R. Pomraenke, G. Cirmi, E. De Re, W. Wang, S. Schwieger, D. Leipold, E. Runge, G. Cerullo, C. Lienau, ACS Nano. **4**, 7559 (2010)
24. T. Schwartz, J.A. Hutchison, C. Genet, T.W. Ebbesen, Phys. Rev. Lett. **106**, 196405 (2011)
25. P. Vasa, W. Wang, R. Pomraenke, M. Lammers, M. Maiuri, C. Manzoni, G. Cerullo, C. Lienau, *Nature Photon.* **7**, 128 (2013) (and supporting on-line material)
26. W. Wang, P. Vasa, R. Pomraenke, R. Vogelgesang, A. de Sio, E. Sommer, M. Maiuri, C. Manzoni, G. Cerullo, C. Lienau, ACS Nano. **8**, 1056 (2014)
27. P. Vasa, W. Wang, R. Pomraenke, M. Maiuri, C. Manzoni, G. Cerullo, C. Lienau, *Phys. Rev. Lett.* **114**, 036802 (2015) (and supporting on-line material)
28. T. Kobayashi, *J aggregates* (World Scientific, Singapore, 1996)
29. H. Raether, *Surface Plasmons on Smooth and Rough Surfaces and on Gratings* (Springer, Berlin, 1988)
30. S.A. Maier, *Plasmonics: Fundamentals and Applications* (Springer Science+Business Media LLC, New York, 2007)
31. W.L. Barnes, A. Dereux, T.W. Ebbesen, Nature **424**, 824 (2003)
32. M.I. Stockman, Opt. Exp. **19**, 22029 (2011)
33. K. Kneipp, Y. Wang, H. Kneipp, L.T. Perelman, I. Itzkan, R.R. Dasari, M.S. Feld, Phys. Rev. Lett. **78**, 1667 (1997)
34. S. Schlücker, Angew. Chem. Int. Ed. **53**, 4756 (2014)
35. U. Akram, Z. Ficek, S. Swain, Phys. Rev. A **62**, 013413 (2000)
36. R.G. DeVoe, R.G. Brewer, Phys. Rev. Lett. **76**, 2049 (1996)
37. C. Ropers, D.J. Park, G. Stibenz, G. Steinmeyer, J. Kim, D.S. Kim, C. Lienau, Phys. Rev. Lett. **94**, 113901 (2005)
38. D.S. Kim, S.C. Hohng, V. Malyarchuk, Y.C. Yoon, Y.H. Ahn, K.J. Yee, J.W. Park, J. Kim, Q.H. Park, C. Lienau, Phys. Rev. Lett. **91**, 143901 (2003)
39. G. Lindblad, Commun. Math. Phys. **48**, 119 (1976)
40. R.H. Dicke, Phys. Rev. **93**, 99 (1954)
41. D.E. Chang, A.S. Sorensen, E.A. Demler, M.D. Lukin, Nature Phys. **3**, 807 (2007)
42. M. Kauranen, A.V. Zayats, Nature Photon. **6**, 737 (2012)
43. R.F. Oulton, V.J. Sorger, T. Zentgraf, R.M. Ma, C. Gladden, L. Dai, G. Bartal, X. Zhang, Nature **461**, 629 (2009)

44. V.I. Klimov, S.A. Ivanov, J. Nanda, M. Achermann, I. Bezel, J.A. McGuire, A. Piryatinski, Nature **447**, 91 (2007)
45. R. Houdré, J.L. Gibernon, P. Pellandini, R.P. Stanley, U. Oesterle, C. Weisbuch, J. O'Gorman, B. Roycroft, M. Ilegems, Phys. Rev. B **52**, 7810 (1995)
46. F. Jahnke, M. Kira, S.W. Koch, G. Khitrova, E.K. Lindmark, T.R. Nelson Jr., D.V. Wick, J.D. Berger, O. Lyngnes, H.M. Gibbs, Phys. Rev. Lett. **77**, 5257 (1996)
47. A. Solomon, C. Genet, T.W. Ebbesen, Angew. Chem. Int. Ed. **48**, 8748 (2009)
48. P. Vasa, C. Lienau, Angew. Chem. Int. Ed. **49**, 2476 (2010)
49. J.A. Hutchison, T. Schwartz, C. Genet, E. Devaux, T.W. Ebbesen, Angew. Chem. Int. Ed. **51**, 1592 (2012)
50. C.K. Dass, T. Jarvis, V.P. Kunets, Y.I. Mazur, G.G. Salamo, C. Lienau, P. Vasa, X. Li, ACS Photonics **2**, 1341 (2015)
51. T. Schwartz, J.A. Hutchison, J. Lèonard, C. Genet, S. Haacke, T.W. Ebbesen, Chem. Phys. Chem. **14**, 125 (2013)
52. D.M. Whittaker, P. Kinsler, T.A. Fisher, M.S. Skolnick, A. Armitage, A.M. Afshar, M.D. Sturge, J.S. Roberts, Phys. Rev. Lett. **77**, 4792 (1996)
53. P. Michetti, G.L. Rocca, Phys. Rev. B **79**, 035325 (2009)
54. A. Hayat, A.C. Lange, L.A. Rozema, A. Darabi, H.M. van Driel, A.M. Steinberg, B. Nelsen, D.W. Snoke, L.N. Pfeiffer, K.W. West, *Phys. Rev. Lett.* **109**, 033605 (2012)
55. B.R. Mollow, Phys. Rev. A **5**, 2217 (1972)
56. H. Häffner, S. Gulde, M. Riebe, G. Lancaster, C. Becher, J. Eschner, F. Schmidt-Kaler, R. Blatt, Phys. Rev. Lett. **90**, 143602 (2003)
57. M. Saba, F. Quochi, C. Ciuti, D. Martin, J.-L. Staehli, B. Deveaud, A. Mura, G. Bongiovanni, Phys. Rev. B **62**, R16322 (2000)
58. T. Unold, K. Mueller, C. Lienau, T. Elsaesser, A.D. Wieck, Phys. Rev. Lett. **92**, 157401 (2004)
59. P. Vasa, C. Ropers, R. Pomraenke, C. Lienau, Laser Photon. Rev. **3**, 483 (2009)
60. J. Zhang, Y. Tang, K. Lee, M. Ouyang, Nature **466**, 91 (2010)
61. J.M. Pirkkalainen, S.U. Cho, J. Li, G.S. Paraoanu, P.J. Hakonen, M.A. Sillanpaa, Nature **494**, 211 (2013)
62. B.B. Buckley, G.D. Fuchs, L.C. Bassett, D.D. Awschalom, Science **330**, 1212 (2010)
63. M. Combescot, R. Combescot, Phys. Rev. Lett. **61**, 117 (1988)
64. W.H. Knox, D.S. Chemla, D.A.B. Miller, J.B. Stark, S. Schmitt-Rink, Phys. Rev. Lett. **62**, 1189 (1989)
65. D.J. Bergman, M.I. Stockman, Phys. Rev. Lett. **90**, 027402 (2003)
66. J. Seidel, S. Grafstroem, L. Eng, Phys. Rev. Lett. **94**, 177401 (2005)
67. H. Deng, G. Weihs, C. Santori, J. Bloch, Y. Yamamoto, Science **298**, 199 (2002)
68. J. Kasprzak, M. Richard, S. Kundermann, A. Baas, P. Jeambrun, J.M.J. Keeling, F.M. Marchetti, M.H. Szymańska, R. André, J.L. Staehli, V. Savona, P.B. Littlewood, B. Deveaud, D. Le Si, Nature **446**, 409 (2006)
69. L. Shi, T.K. Hakala, H.T. Rekola, J.-P. Martikainen, R.J. Moerland, P. Törmä, Phys. Rev. Lett. **112**, 153002 (2014)
70. G. Zengin, M. Wersäll, S. Nilsson, T.J. Antosiewicz, M. Käll, T. Shegai, Phys. Rev. Lett. **114**, 157401 (2015)
71. K. Santhosh, O. Bitton, L. Chuntonov, G. Haran, Nat. Commun. **7**, 11823 (2016)
72. R. Chikkaraddy, B. de Nijs, F. Benz, S.J. Barrow, O.A. Scherman, E. Rosta, A. Demetriadou, P. Fox, O. Hess, J.J. Baumberg, Nature **7**, 11823 (2016)
73. R. Liu, Z.-K. Zhou, Y.-C. Yu, T. Zhang, H. Wang, G. Liu, Y. Wei, H. Chen, X.-H. Wang, Phys. Rev. Lett. **118**, 237401 (2017)
74. A. Gonzalez-Tudela, D. Martin-Cano, E. Moreno, L. Martin-Moreno, C. Tejedor, F.J. Garcia-Vidal, Phys. Rev. Lett. **106**, 020501 (2011)
75. A. Emboras, J. Niegemann, P. Ma, C. Haffner, A. Pedersen, M. Luisier, C. Hafner, T. Schimmel, J. Leuthold, Nano Lett. **16**, 709 (2016)

Chapter 6
Coherent Control of Nonadiabatic Dynamics of Electron-Phonon Systems by Quantized Light Field

Kunio Ishida

Abstract Coherent control of electron-phonon systems is studied by numerical calculations on a toy model. We found that the interplay of the electron-phonon nonadiabaticity and the Raman scattering determines the wavepacket motion particularly in the vicinity of the conical intersection of adiabatic potential energy surfaces. Entanglement between photons, phonons, or electrons is discussed by calculating the entanglement entropy. We showed that the quantum nature of irradiated light plays an important role in considering design and/or manipulation methods of the quantum-mechanical states of electron-phonon systems in coherent regime.

6.1 Introduction

Coherent control of quantum-mechanical states have been studied since the proposal of the adaptive control method [1], which are considered to be promising for innovative technology particularly in the field of quantum information. For example, recent experiments on diamond showed the possibility of phonon-mediated coherence between qubits via Raman scattering processes [2]. In this case, entanglement between electrons, phonons, and photons plays an important role, which means that theory of quantum entanglement between irradiated light and materials is necessary to reveal and/or design the control methods for them. In other words, we should study the quantum dynamics of the full Hamiltonian regarding all the above entities as quantum variables, taking into account the interplay among electrons, photons, and phonons.

On the other hand, recent progress of ultrashort pulse laser technology has made it possible to observe time evolution of quantum-mechanical states in coherent regime. Above all, time-resolved techniques have been developed to clarify the excited-state dynamics of condensed matter, and its transient properties have been studied [3–6].

K. Ishida (✉)
School of Engineering, Utsunomiya University, 7-1-2 Yoto,
Utsunomiya, Tochigi 321-8585, Japan
e-mail: ishd_kn@cc.utsunomiya-u.ac.jp

K. Yamanouchi and D. Charalambidis (eds.), *Progress in Ultrafast Intense Laser Science XV*, Topics in Applied Physics 136,
https://doi.org/10.1007/978-3-030-47098-2_6

When, for example, ultrashort laser pulses are used as a probe, transient electronic properties are detected through the slight change of optical properties, which helps us understand the physics in ultrafast timescale.

Another example of the interesting transient phenomena is the structural dynamics. Time-resolved x-ray or electron diffraction measurement has been intensively developed, and the dynamics of photoinduced structural change or chemical reaction is currently being discussed by many authors [7, 8]. Furthermore, as various experimental methods have been developed to understand transient phenomena [9, 10], it is quite important to combine as many experimental data as possible in order to obtain an overall picture for the transient dynamics of certain materials. We, however, should point out that we need a unified physical model to understand all the experimental data on a single phenomenon. In particular, when we are interested in time evolution of excited states, we consider that a first step to construct an appropriate model is an accurate description of photoexcitation/deexcitation processes.

Considering that electromagnetic field has been regarded as classical external field in many theoretical studies, we study the quantum dynamics of electronic systems under photoirradiation, focusing on the wavepacket states created by photons. We consider that we will have a reliable information on the response to various probes and the entanglement properties between multiple degrees of freedom by determining the quantum nature of those created wavepackets. Since, however, it is not straightforward to obtain a first-principles theory for transient dynamics, we focus on the interplay of electron-phonon interaction and electron-photon (electric dipole) interaction as a first step to construct a general theory of transient photoexcitation/deexciation phenomena. For this purpose we chose a model of coupled electron-phonon-photon systems and solved the time-dependent Schrödinger equation for fully quantized systems in order to discuss the wavepacket motion in the presence of electromagnetic field.

We also mention that nonadiabatic coupling between potential energy surfaces (PESs) is a key to understand the relaxation dynamics of photoexcited states, e.g., photoinduced nucleation [11]. Furthermore, when electrons, phonons, and photons are considered at the same time, the Raman processes are expected to give significant contribution to the electronic transitions, which means that the photoexcitation/deexcitation process of electron-phonon systems should be carefully dealt with in order to understand the ultrafast dynamics with light-matter interaction. As well as the intermodal coherence mediated by the Raman processes [12], we stress that it is worth while mentioning that previous studies [13–16] have shown that the conical intersection (CI) in the "classical" adiabatic PESs also is a key to understand the wavepacket dynamics. These results show us that the coexistence of nonadiabatic coupling between PESs and Raman scattering processes gives another viewpoint on the coherent dynamics of electron-phonon-photon systems. The aim of the present paper is to study the interplay of electron-phonon nonadiabaticity and Raman scattering processes by numerical calculations with a toy model.

6.2 Model and Method

In this paper we study the quantum dynamics of electrons coupled with both phonons and photons taking into account the Raman scattering processes. For this purpose, we employed a model of a two-level electronic system coupled with a single-mode phonons and multimode photons described by

$$\mathcal{H} = \omega a^\dagger a + \sum_{i=1}^{n} \Omega_i c_i^\dagger c_i + \sigma_x \left\{ \sum_{i=1}^{n} \mu_i (c_i^\dagger + c_i) + \lambda \right\} + \frac{1}{2}(\sigma_z + 1)\{\nu(a^\dagger + a) + \varepsilon\}, \tag{6.1}$$

where a and c_i denote the annihilation operators of phonons and the photons of the i-th mode, respectively. σ_i corresponds to the Pauli matrices which operate on the electronic states denoted by $|g\rangle$ (ground state) and $|e\rangle$ (excited state). We should refer to the Jaynes-Cummings model [17] which formally includes the same type of interaction between electrons and bosons as (6.1), although we consider two kinds of bosons, i.e., phonons and photons.

Corresponding PESs of the Hamiltonian (6.1) are obtained by regarding the amplitude operators $\hat{u} = (a^\dagger + a)/\sqrt{2\omega}$, $\hat{v}_i = (c_i^\dagger + c_i)/\sqrt{2\Omega_i}$ as classical variables. Although we do not consider any classical motion on those PESs, it still helps us understand the overall behavior of the wavepackets by discussing the adiabatic PESs which are given by

$$U_\pm(u, v_1, v_2, \ldots, v_n) = \frac{\omega^2}{2}u^2 + \sum_{i=1}^{n} \frac{\Omega_i^2}{2} v_i^2 + \frac{1}{2}(\nu' u + \varepsilon) \pm \sqrt{(\nu u' + \varepsilon)^2 + \left(\sum_{i=1}^{n} \mu_i' v_i + \lambda\right)^2}, \tag{6.2}$$

where $\nu' = \sqrt{2\omega}\nu$ and $\mu_i' = \sqrt{2\Omega_i}\mu_i$. Equation (6.2) shows that the adiabatic PESs have a CI given by $u = -\varepsilon/\nu'$ and $\sum_i^n \mu_i' v_i = -\lambda$, and thus the geometrical phase of the wavefunction plays an important role on the dynamical properties [15].

The time-dependent Schrödinger equation for Hamiltonian (6.1) is numerically solved for $n = 1$ and 3 and to obtain the wavefunction $|\Phi(t)\rangle$. The initial condition is given by $|\Phi(0)\rangle = |\alpha_1, \alpha_2, \ldots, \alpha_n\rangle \otimes |0g\rangle$, where $|\alpha_i\rangle$ denotes a coherent state parameterized by α_i for the i-th photon mode and $|0g\rangle$ is the ground state of the electron-phonon system. The values of the parameters are $\omega = 1$, $\mu_i = 0.5 (i = 1, 2, 3)$, $\nu = 3.5$, and $\varepsilon = 13.5$, which shows that the electron-phonon coupling (Huang-Rhys factor) has intermediate strength between solid [18] and typical organic molecules [19]. As for the photons, we consider cases with a single mode ($n = 1$) and three modes ($n = 3$), where $\alpha_1 = 5$ in all cases. Mode 1 (pump mode) which is in resonance with the Franck-Condon transition is treated in both cases

($\Omega_1 = 13.5$). Mode 2 (Stokes mode; $\Omega_2 = 12.5$) and mode 3 (anti-Stokes mode; $\Omega_3 = 14.5$) are taken into account in the latter case.

While the internal vibration mode of the material system is always quantized in the present study, we also calculated the dynamics of the material system treating the electromagnetic field as a classical external field for reference, and call this method a semiclassical approximation in the rest of the paper.

6.3 Calculated Results

Since our interest lies in the dynamics of the electron-phonon-photon systems, we first discuss the "diagonal" matrix elements of relevant physical properties, i.e., population of the electronic ground state $N(t) = \langle\Phi(t)|(1+\sigma_z)/2|\Phi(t)\rangle$ and photon number in each mode $n_i(t) = \langle\Phi(t)|c_i^\dagger c_i|\Phi(t)\rangle$.

The solid red line in Fig. 6.1a shows $N(t)$ for $n = 3$ and $\alpha_2 = \alpha_3 = 0$. As the absorption of photon proceeds, $N(t)$ decreases with a rapid oscillation in $t \leq 2$, which is a reminiscent of the Rabi oscillation diminishing with the lattice relaxation.

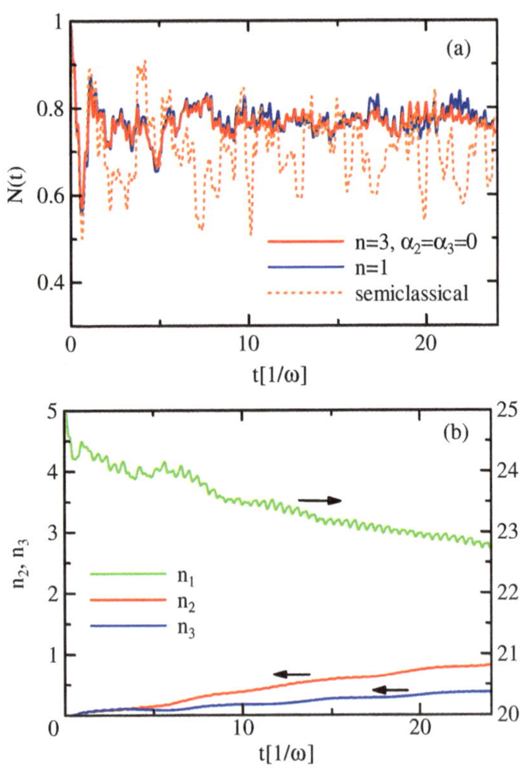

Fig. 6.1 **a** The ground state population of electron $N(t)$ for $\lambda = 1.5$ as functions of time. The solid red line shows the result for $n = 3$ and $\alpha_2 = \alpha_3 = 0$, and the blue line shows that for $n = 1$. The dotted line is $N(t)$ by the semiclassical approximation for $n = 3$. **b** Photon number n_i $(i = 1, 2, 3)$ for $n = 3$ and $\alpha_2 = \alpha_3 = 0$

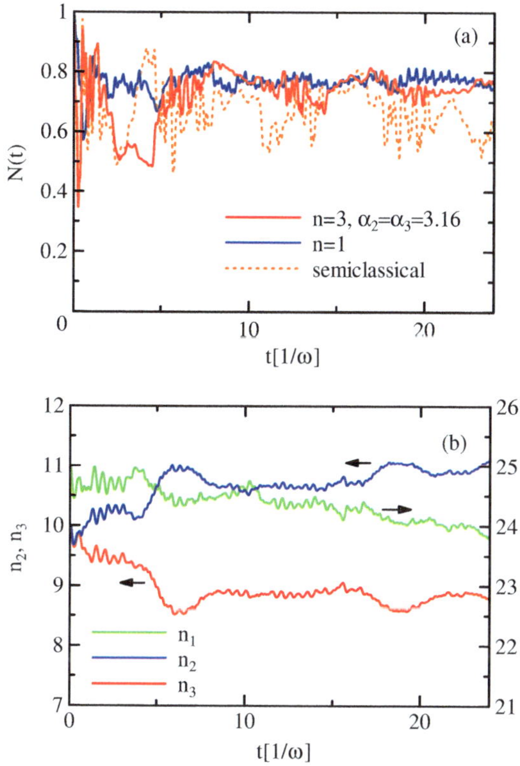

Fig. 6.2 **a** The ground state population of electron $N(t)$ for $\lambda = 1.5$ as functions of time. The solid line shows the quantum dynamics of $N(t)$ for $n = 3$ and $\alpha_2 = \alpha_3 = 3.16$, and the dotted line shows the result obtained by the semiclassical approximation for $n = 3$. **b** Photon number n_i $(i = 1, 2, 3)$ for $n = 3$ and $\alpha_2 = \alpha_3 = 3.16$

When only a single-mode photon is taken into account, i.e., $n = 1$, $N(t)$ behaves similarly to that for $n = 3$, which shows that the Raman processes plays a minor role in the wavepacket dynamics. This interpretation is also supported by the behavior of n_2 and n_3 shown in Fig. 6.1b. This figure shows that the increase of the photon number for modes 2 and 3 is small even when the wavepacket motion proceeds and the energy difference between two corresponding adiabatic PESs is resonant to the Stokes mode.

The solid red line in Fig. 6.2a shows $N(t)$ for $\alpha_2 = \alpha_3 = 3.16$ and $\lambda = 1.5$, and the dotted line is a corresponding property calculated by the semiclassical approximation, where both the Stokes and the anti-Stokes modes have finite intensity at $t = 0$. Although semiclassical approximation is good for $t < 2$, deviation between them increases rapidly thereafter. Accordingly, as Fig. 6.2b shows, photons in mode 2 and 3 increase and decrease, respectively, which shows that the Raman processes contribute to the dynamics of the whole system. To be more precise, n_2 and n_3 rapidly change their value at $t \sim 2\pi$, and the Raman processes modulate electronic transition, which shows that the stimulated Raman process enhances the electronic transition for $\alpha_2, \alpha_3 \neq 0$.

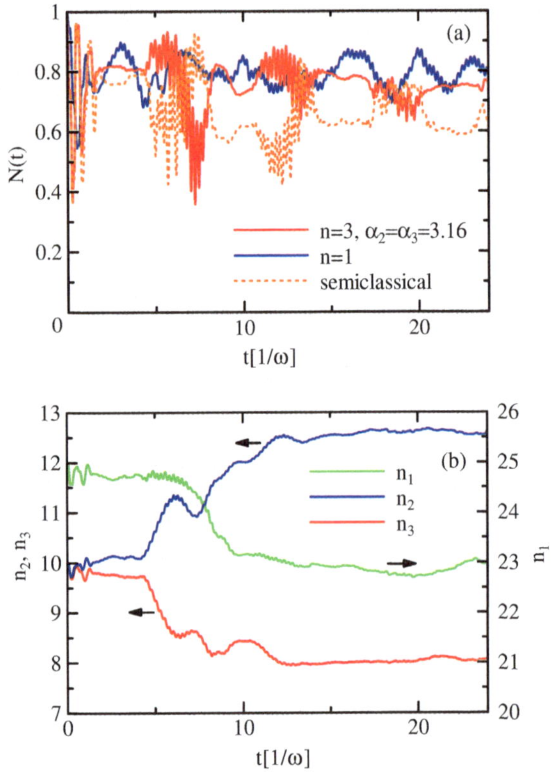

Fig. 6.3 **a** The ground state population of electron $N(t)$ for $\lambda = 0$ as functions of time. The solid line shows the quantum dynamics of $N(t)$ for $n = 3$ and $\alpha_2 = \alpha_3 = 3.16$, and the dotted line shows the result obtained by the semiclassical approximation for $n = 3$. **b** Photon number n_i $(i = 1, 2, 3)$ for $n = 3$ and $\alpha_2 = \alpha_3 = 0$

Comparing the temporal behavior of n_1 in Figs. 6.1b and 6.2b, we found that the absorption of the pump-mode photons is suppressed when both the Stokes and the anti-Stokes Raman processes take place. Since these processes are relevant to both the absorption and the emission of the pump-mode photons, we consider that the interference between two Raman processes contributes to the dynamics of the whole system. This effect does not disturb the transition between $|g\rangle$ and $|e\rangle$ and thus $N(t)$ changes its value as Fig. 6.1a shows. Since such processes are not taken into account for $n = 1$, the difference between the solid red line and the blue line in Fig. 6.2a is much larger than that in Fig. 6.1a.

The features discussed above are understood clearly by comparing the above results to those without the electron-phonon nonadiabatic coupling ($\lambda = 0$). Figure 6.3a, b show $N(t)$ and $n_i(t)$ for $\alpha_2 = \alpha_3 = 3.16$ and $\lambda = 0$. First, we point out that n_2/n_3 increases/decreases stepwise around $t = 2\pi$ and 4π. Furthermore, absorption of mode 2 and emission of mode 3 synchronously take place, while photons of mode 1 are absorbed subsequently. Since, however, no two-phonon/photon process is allowed to the lowest order, we point out that the nonlinear behavior of n_2/n_3 is caused by the resonance of the Stokes and the anti-Stokes Raman processes at this stage. Then, after the Raman processes become off-resonant, the interference between them turns to be weaker and the absorption of mode 1 photons is clearly seen.

Comparing Figs. 6.2a with 6.3a, we found that electronic transition is suppressed at $t \sim 2\pi$ and 4π also by the electron-phonon nonadiabaticity. We note that the adiabatic PESs of the electron-phonon subsystem (PESSs) have an avoided crossing at $u = -\varepsilon/\nu'$ for $\lambda \neq 0$. The photoexcited wavepackets bifurcate at the avoided-crossing and thus the transition between $|g\rangle$ and $|e\rangle$ is more complicated for finite λ. In this case, the interference between those processes affects the electronic transition, and the resonance to the pump mode photons is blurred. Hence, the temporal change of $N(t)$ and $n_i(t)$ for $t \sim 10$ is unclear as shown in Fig. 6.2a, b.

Since the wavepacket trajectory on the PESSs is an experimentally observable quantity [20], we calculated the lattice displacement $u(t) = \langle\Phi(t)|\hat{u}|\Phi(t)\rangle$ and focus on the dynamics of the electron and phonons. Figure 6.4a–c show $u(t)$ for the three cases corresponding to Figs. 6.1, 6.2 and 6.3. As discussed previously, Fig. 6.4a, b show that the semiclassical approximation is not valid for $t > 2\pi$ for $\lambda \neq 0$. On the contrary, the solid line and the dotted line in Fig. 6.4c are similar to each other, which shows that the nonadiabaticity of the electron-phonon-photon dynamics is relevant to the validity of the semiclassical approximation. As mentioned above, the (one-dimensional) PESSs have an avoided-crossing at $u = -\varepsilon/\nu'$, while the PESs for the whole system has a CI. Since the semiclassical approximation takes into account only the avoided-crossing, the wavepacket motion bifurcate in a different manner between the quantum-mechanical calculation and the semiclassical calculation particularly in the vicinity of the avoided-crossing or the CI, which results in the different dynamics or trajectory shown in the figures. We stress that the quantum-mechanical nature of the incident light plays an important role on the wavepacket motion, and that detailed discussion will be possible by revealing the transient dynamics of coherent phonons by ultrafast optical spectroscopy. To be more precise, the role of the CI on the electronic transition should be revealed in order to determine the wavepackets created by photons. In particular, as the irradiation of photons proceeds, deviation from single-mode model becomes larger, the role of CI becomes more important.

As an example of coherent control by external light field, we show that the relative phase between photons affects the created phonon states. Figure 6.5 shows the number of phonons $N_n(t) = \langle\Phi(t)|a^\dagger a|\Phi(t)\rangle$ for $\Delta\phi = -\pi, -\pi/2, 0, \pi/2$ and π, where $\Delta\phi$ denotes the phase difference between Stokes/anti-Stokes mode and the pump mode, i.e., $\alpha_2 = |\alpha_2|e^{i\Delta\phi}$ and $\alpha_3 = |\alpha_3|e^{-i\Delta\phi}$. Since the wavepacket trajectory varies with $\Delta\phi$, different phonon states are created in the material, which shows that the number of phonons is able to be modified by $\Delta\phi$. These results show that, when we consider the control of phonon states by external light, not only the duration of the photon pulses but also the relative phase between photon modes is important.

The quantum-mechanical nature of the electromagnetic field is reflected on the entanglement between subsystems, i.e., photons, phonons, and electrons. Hence, we calculated the bipartite entanglement entropy in which the whole system is divided into (i) the electronic system and the phonon-photon system, and (ii) the electron-phonon subsystem and the photons. Although it is not directly observable in experiments, we will have an important information on the quantum-mechanical nature

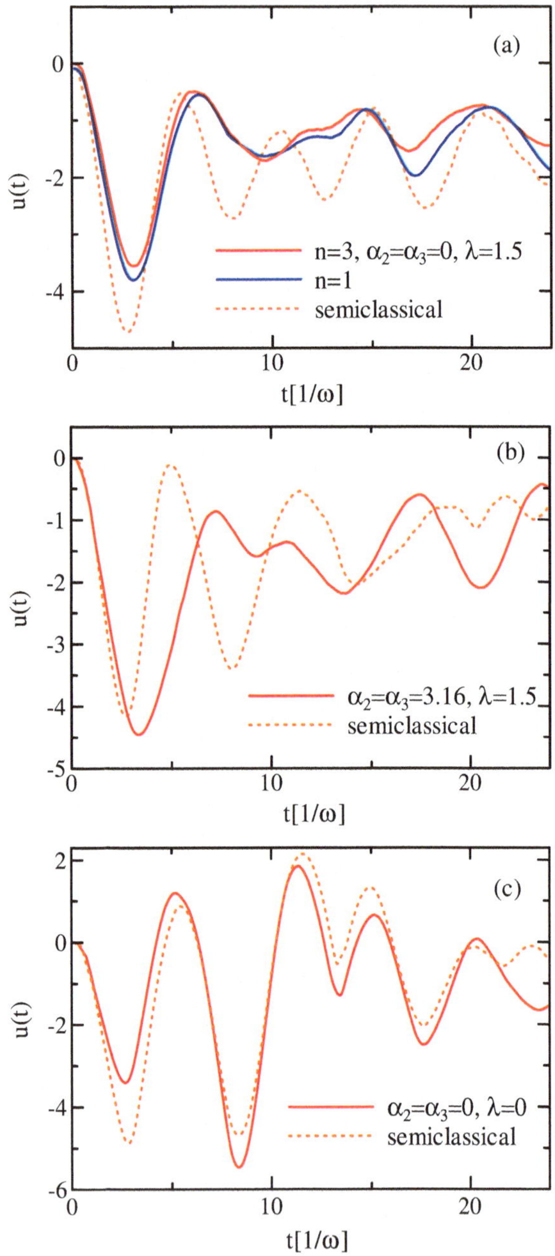

Fig. 6.4 Lattice displacement $u(t)$ for **a** $\alpha_2 = \alpha_3 = 0,\ \lambda = 1.5$, **b** $\alpha_2 = \alpha_3 = 3.16,\ \lambda = 1.5$, and **a** $\alpha_2 = \alpha_3 = 3.16,\ \lambda = 0$. Calculated results with the semiclassical approximation are shown by the dotted line, and the blue line in (**a**) shows $u(t)$ for $n = 1$

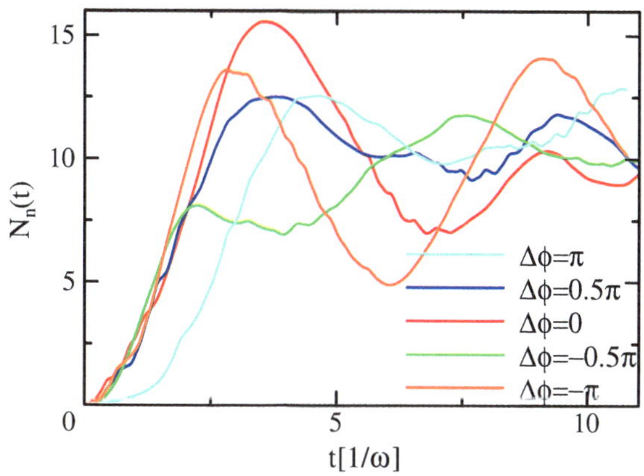

Fig. 6.5 The number of phonons $N_n(t)$ for $\alpha_2 = \alpha_3 = 3.16$, $\lambda = 1.5$ with different values of $\Delta\phi$

of the wavepackets created in the material system. The entropy for these cases is defined by

$$S_a(t) = \mathrm{Tr}\rho_a(t) \log \rho_a(t), \tag{6.3}$$

$$S_b(t) = \mathrm{Tr}\rho_b(t) \log \rho_b(t), \tag{6.4}$$

where

$$\rho_a(t) = \mathrm{Tr}_{pn,pt}|\Phi(t)\rangle\langle\Phi(t)|, \tag{6.5}$$

$$\rho_b(t) = \mathrm{Tr}_{pt}|\Phi(t)\rangle\langle\Phi(t)|. \tag{6.6}$$

$\mathrm{Tr}_{pn,pt}$ and Tr_{pt} denote the partial trace of the density matrix regarding the phonon and photon degrees of freedom, and the photon degrees of freedom, respectively. We point out that S_a is the entropy of a two-level system and that its value lies between 0 and log 2 $\sim$0.693.

Figure 6.6a–c show that entanglement represented by (i) and (ii) grows immediately after the simulation starts. Figure 6.6b, c also show that S_a and S_b have a fine structure corresponding to the increase/decrease of n_2/n_3, i.e., the Raman processes enhance the rate of entropy production. As for the effect of the electron-phonon nonadiabaticity, we point out that Fig. 6.6c shows that the photoabsorption at $t \sim 2\pi$ and 4π decreases S_b, which shows that the coherence between the electronic states recovers by the external field, i.e., photons. As shown in the other properties, the interference mechanism between electronic states becomes different in the presence of the electron-phonon nonadiabaticity and the decrease of S_b is not observed in Fig. 6.6a or b.

Thus, although the quantum coherence can be controlled by by irradiated photons, we should note that entanglement between photons and phonons is also disturbed by

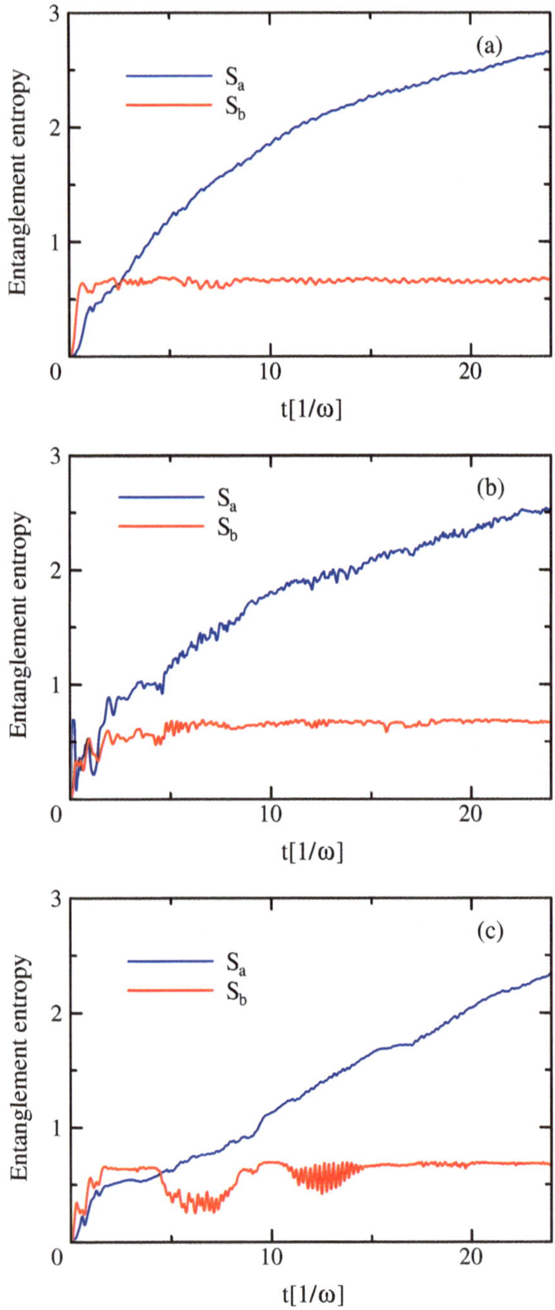

Fig. 6.6 Entanglement entropy $S_a(t)$ and $S_b(t)$ for **a** $\alpha_2 = \alpha_3 = 0,\ \lambda = 1.5$, **b** $\alpha_2 = \alpha_3 = 3.16,\ \lambda = 1.5$, and **c** $\alpha_2 = \alpha_3 = 3.16,\ \lambda = 0$

their interaction with electrons through μ_i and ν. In order to clarify the mechanism of this effect, further discussion on the entanglement entropy in multipartite systems [21–23] will be helpful.

6.4 Summary

We calculated the wavepacket dynamics of electron-phonon-photon systems by numerical calculations focusing on the interplay of the electron-phonon nonadiabaticity and the Raman processes. While the effect of the Raman processes is small on the wavepacket motion starting with empty Stokes/anti-Stokes modes, the stimulated Raman processes enhance the electronic transition, which shows that we should take into account them to obtain accurate trajectories of wavepackets. In particular, the interference between Stokes and anti-Stokes Raman processes suppresses the absorption of pump-mode photons, although the electronic excitation is induced and the excited-state population $N(t)$ takes a large value. Furthermore, when both λ and ν_i are present, large deviation from semiclassical calculations is obtained, which shows that the bifurcation of wavepackets should be carefully treated in the presence of both the electron-phonon adiabaticity and the Raman processes.

As an example of coherent control of phonon states by external light, we show that the relative phase between photon modes is directly reflected to the created phonon states. However, the role of the CI on the phonon creation mechanism is still an open question, and we need further study on the modulation of the wavepacket trajectory which induces the change of the winding number around the CI [15], for example.

References

1. R.S. Judson, H. Rabitz, Phys. Rev. Lett. **68**, 1500 (1992)
2. D.G. England, K.A.G. Fisher, J.W. MacLean, P.J. Bustard, K. Heshami, K.J. Resch, B.J. Sussman, Phys. Rev. Lett. **117**, 073603 (2016)
3. R.M. Bowman, M. Dantus, A.H. Zewail, Chem. Phys. Lett. **174**, 546 (1990)
4. K.G. Nakamura, K. Ohya, H. Takahashi, T. Tsuruta, H. Sasaki, S. Uozumi, K. Norimatsu, M. Kitajima, Y. Shikano, Y. Kayanuma, Phys. Rev. B **94**, 024303 (2016)
5. M. Gallart, M. Ziegler, O. Crégut, E. Feltin, J.-F. Carlin, R. Butté, N. Grandjean, B. Hönerlage, P. Gilliot, Phys. Rev. **B96**, 041303 (2017)
6. T. Ghosh, S. Aharon, A. Shpatz, L. Etgar, S. Ruhman, ACS Nano. **12**, 5719 (2018)
7. M. Gao, C. Lu, H. Jean-Ruel, L.C. Liu, A. Marx, K. Onda, S. Koshihara, Y. Nakano, X. Shao, T. Hiramatsu, G. Saito, H. Yamochi, R.R. Cooney, G. Moriena, G. Sciaini, R.J.D. Miller, Nature **496**, 343 (2013)
8. K.H. Kim, J.G. Kim, S. Nozawa, T. Sato, K.Y. Oang, T.W. Kim, H. Ki, J. Jo, S. Park, C. Song, T. Sato, K. Ogawa, T. Togashi, K. Tono, M. Yabashi, T. Ishikawa, J. Kim, R. Ryoo, J. Kim, H. Ihee, S. Adachi, Nature **518**, 385 (2015)
9. M. Först, A.D. Caviglia, R. Scherwitzl, R. Mankowsky, P. Zubko, V. Khanna, H. Bromberger, S.B. Wilkins, Y.-D. Chuang, W.S. Lee, W.F. Schlotter, J.J. Turner, G.L. Dakovski, M.P. Minitti,

J. Robinson, S.R. Clark, D. Jaksch, J.-M. Triscone, J.P. Hill, S.S. Dhesi, A. Cavalleri, Nat. Mater. **14**, 883 (2015)
10. S. Yoshida, Y. Aizawa, Z. Wang, R. Oshima, Y. Mera, E. Matsuyama, H. Oigawa, O. Takeuchi, H. Shigekawa, Nat. Nanotech. **9**, 588 (2014)
11. K. Ishida, K. Nasu, Phys. Rev. Lett. **100**, (2008)
12. B. Sen, S. Kumar Giri, S. Mandal, C.H. Raymond Ooi, A. Pathak, Phys. Rev. **A87**, 022325 (2013)
13. M. Kowalewski, K. Bennett, S. Mukamel, J. Chem. Phys. **144**, 05439 (2016)
14. S.C. Althorpe, J. Chem. Phys. **103**, 084105 (2006)
15. F. Bouakline, Chem. Phys. **442**, 31 (2014)
16. C. Xie, D.R. Yarkony, H. Guo, Phys. Rev. A **95**, 022104 (2017)
17. E.T. Jaynes, F.W. Cummings, Proc. IEEE **51**, 89 (1963)
18. A. Hoffmann, B.K. Meyer, E. Malguth, *Zinc Oxide: From Fundamental Properties Towards Novel Applications*, vol. 120 (Springer, Heidelberg, 2010), p. 233
19. R. Valiev, A. Freidzon, A. Berezhnoy, RSC Adv. **4**, 79 (2014)
20. K. Ishida, F. Aiga, K. Misawa, J. Chem. Phys. **127**, 194304 (2007)
21. C. Moura Alves, D. Jaksch, Phys. Rev. Lett. **93**, 110501 (2004)
22. S. Gerke, J. Sperling, W. Vogel, Y. Cai, J. Roslund, N. Treps, C. Fabre, Phys. Rev. Lett. **117**, 110502 (2016)
23. F. Clivaz, M. Huber, L. Lami, G. Murta, J. Math. Phys. **58**, 082201 (2017)

Chapter 7
Application of CR-39 Solid State Nuclear Track Detectors to Laser-Driven Ion Acceleration Experiments

Masato Kanasaki, Tomoya Yamauchi, Keiji Oda, and Yuji Fukuda

Abstract The solid state nuclear track detector CR-39 is the most reliable ion detector used in laser-driven ion acceleration experiments. This is because CR-39 detectors can measure from *one* ion without X-ray and electron noise generated by the interaction between the intense laser pulse and the target matter. In addition, the incident energy and the spatial distributions of the laser-accelerated ions are also revealed by the careful analysis of the etch pits in CR-39 detectors. The present paper comments on CR-39 detectors based on previous studies with a focus on an analysis technique of the etch pits to obtain the incident energy of the laser-accelerated ions.

7.1 Introduction

Laser-driven ion acceleration has been one of the most active areas of research in the last several years, given that accelerated ions have unique properties, such as ultrashort duration, high brilliance, and low emittance [1, 2]. Recently, protons with energies exceeding 94 MeV were recorded via a hybrid scheme of radiation pressure-sheath acceleration in an ultrathin foil irradiated by a linearly polarised laser pulse [3]. Here, understanding the mechanism of the ion acceleration is critical for improving the energy and quality of laser-accelerated ion beams. A precise quantitative characterization of ion beams in the both space and energy domains allows us to discuss the physics of the laser-driven ion acceleration process.

In such a situation, CR-39 detectors, which are the most sensitive solid state nuclear track detectors (SSNTDs), have been used as the most reliable detector for

M. Kanasaki · T. Yamauchi · K. Oda
Graduate School of Maritime Sciences, Kobe University, 5-1-1 Fukaeminami-machi, Higashinada-ku, Kobe 658-0022, Japan
e-mail: kanasaki@maritime.kobe-u.ac.jp

Y. Fukuda (✉)
Kansai Photon Science Institute, National Institutes for Quantum and Radiological Science and Technology, 8-1-7 Umemi-dai, Kizugawa-shi, Kyoto 619-0216, Japan
e-mail: fukuda.yuji@qst.go.jp

K. Yamanouchi and D. Charalambidis (eds.), *Progress in Ultrafast Intense Laser Science XV*, Topics in Applied Physics 136,
https://doi.org/10.1007/978-3-030-47098-2_7

a long time in laser-driven ion acceleration experiments [4–6]. This is because CR-39 detectors have great advantages of being insensitive to high-energy photons and electrons at absorbed doses less than approximately 10 kGy [7] and are capable of detecting only ions in mixed radiation fields, such as in laser-driven ion acceleration experiments. In addition, the CR-39 detector (hereafter, referred to as CR-39) can measure the range of each ion by using a multi-step etching technique [8–10]. Recently, this technique was applied to laser-driven ion acceleration experiments to demonstrate a 10-fold improvement in accuracy for determining the maximum ion energy of laser-accelerated protons with an uncertainty of $\Delta E = 0.1$ MeV [11]. CR-39 detectors can also measure the number of ions and spatial distribution of ions unless the etch pit overlap. For example, CR-39 detectors were also used to calibrate the number of ions in a Thomson parabola spectrometer [12, 13].

The present paper comments on the characteristics of SSNTDs and the method of using CR-39 as an ion detector. In particular, an analysis of the etch pits in CR-39 detectors is described in detail to obtain the incident energy of laser-accelerated ions. The application of CR-39 detectors to measurements of laser-accelerated ions from cluster-gas targets is also introduced as an example based on previously published papers by the authors' group.

7.2 Fundamentals of the SSNTDs

7.2.1 Introduction to CR-39 Detectors

CR-39 is usually used as a SSNTD in various fields of research, such as nuclear physics, neutron dosimetry, cosmic ray detection and laser-driven ion acceleration. CR-39 was first developed as an optical material by the Colombia Chemical Division in the 1940s. CR-39 is a colourless and transparent plastic of a polymerized diethylene glycol bis(allyl carbonate) liquid monomer. Figure 7.1 shows the repeated unit of CR-39. The molecular formula is $C_{12}H_{18}O_7$ and the density is 1.31 g/cm^3. CR-39 has two carbonate ester bonds on both sides of the ether bond in the repeat unit. The carbonate ester and the ether bond are well known as radiosensitive groups [14].

In 1978, CR-39 was discovered as a superior SSNTD material by Cartwright et al. [15]. Polyethylene terephthalate (PET), polycarbonate (PC), and cellulose nitrate (CN) were used as SSNTDs before the discover of CR-39 detectors [16–18]. These SSNTD materials exhibit lower sensitivity than CR-39 detectors; for example, CN can detect α-particles with an energy below a few MeV (stopping power of more than

Fig. 7.1 The repeat unit of CR-39

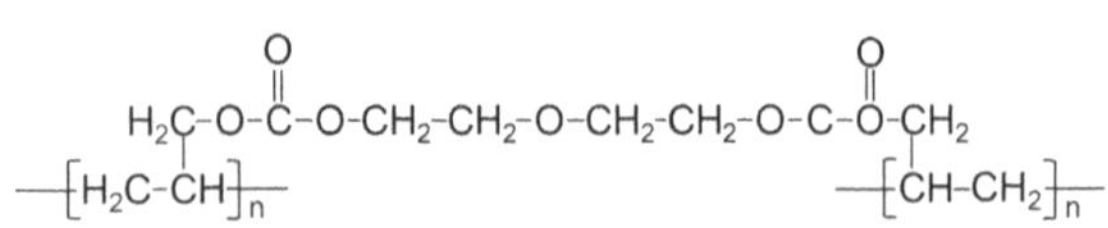

100 keV/μm in the material). On the other hand, CR-39 produced with purified allyl diglycol carbonate monomer can detect protons of a few MeV. In addition, a more sensitive type of CR-39 detector was developed in Japan. For example, HARZLAS (TD-1) and HARZLAS (TNF-1), which are copolymers with N-isopropylacrylamide (NIPAAm), can detect protons up to 20 and 27 MeV (stopping power below 10 keV/μm), respectively [19–21].

The following sections explain the etch pit growth behaviour and analysis of the SSNTDs, including the CR-39, which has been studied by pioneering researchers [8, 22, 23].

7.2.2 *Etch Pit Growth Behaviour*

When an energetic ion enters the SSNTD, the ion transfers kinetic energy to the material and loses energy. Primary ionization by the incident particles and ionization by secondary electrons, which are called δ-rays, mainly occurs in the process of losing energy. Based on these processes, the damaged region, which is called the ion track, is formed with a size of a few nm in the radial direction. The ion track region is preferentially etched by chemical etching compared to the pristine region, and then the etch pit is created. After chemical etching by alkaline solutions, the damaged region can be recognized as an etch pit with an optical microscope. Each etch pit shows that the unique growth behaviour depends on the stopping power along the ion track, which allows us to predict the ion species and the incident energy.

The sensitivity of an SSNTD for a charged particle is defined as the ratio of the track etching speed V_t to the bulk etching speed V_b as described in (7.1).

$$S \equiv \frac{V_t}{V_b}, \tag{7.1}$$

where S is the sensitivity of the SSNTD. V_t and V_b are the etching speeds along the ion track and its unirradiated area, respectively. A schematic cross-sectional view of an etch pit after t hours of etching is shown in Fig. 7.2. The surface of the SSNTD is etched in accordance with V_t and V_b; therefore, the track length L and the thickness of layer removed G are given by $V_t t$ and $V_b t$, respectively. In the case of V_t being constant, the etch pit wall has a perfect conical shape. The relation between the etch pit tip angle δ and the sensitivity is shown in (7.2).

$$\sin\delta = \frac{V_b}{V_t} = \frac{1}{S} \tag{7.2}$$

From (7.2), the etch pit tip angle can determine the sensitivity S directly, but the tip angle cannot be obtained by optical microscopy from the surface of the SSNTD.

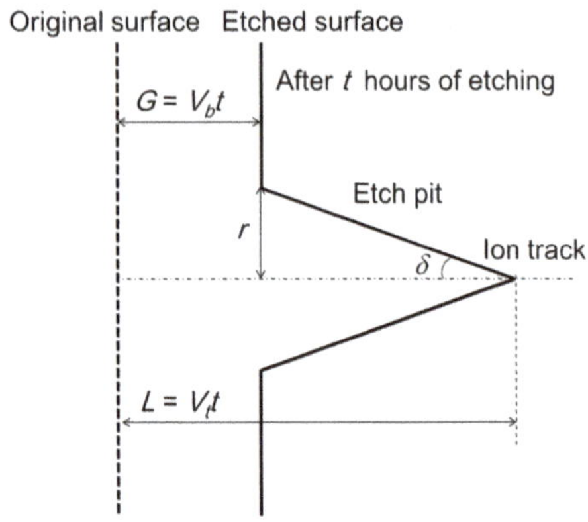

Fig. 7.2 Schematic cross-sectional view of an etch pit after t hours of etching, which assumes constant V_t and V_b

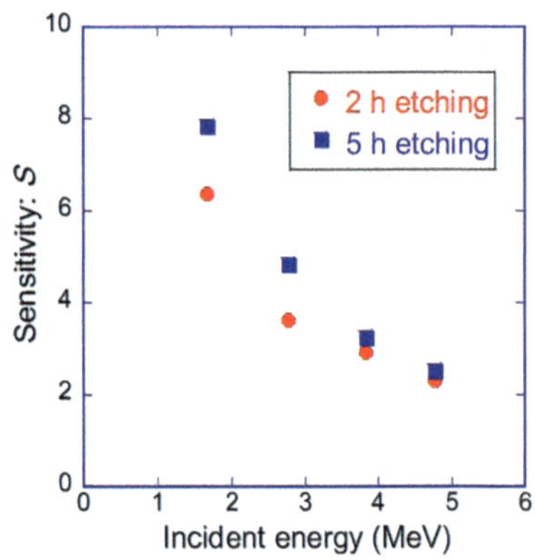

Fig. 7.3 The sensitivity of the CR-39 detectors calculated with (7.3) for α particles with etching times of 2 and 5 h. A longer etching duration shows a higher sensitivity at the same incident energies

Therefore, the usual sensitivity for each etch pit is calculated from the radius of the etch pit open mouth, i.e., the etch pit radius, r, and the thickness of the layer removed G as expressed in (7.3).

$$S = \frac{1 + \left(\frac{r}{G}\right)^2}{1 - \left(\frac{r}{G}\right)^2} \tag{7.3}$$

Here, (7.3) assumes that the V_t and V_b are constant; in other words, the sensitivity does not depend on depth in the detector material, i.e., the etching duration. V_b is always constant in a typical chemical etching treatment. The track etching speed V_t, however, varies along the ion track. This is because ions deposit their energy in accordance with the stopping power; namely, the deposited energy increases towards the end of their range in the detector. For example, Fig. 7.3 shows the sensitivity of CR-39 detectors for α-particles using (7.3) with etching times of 2 and 5 h. Higher sensitivities were observed with longer etching duration at the same incident energies.

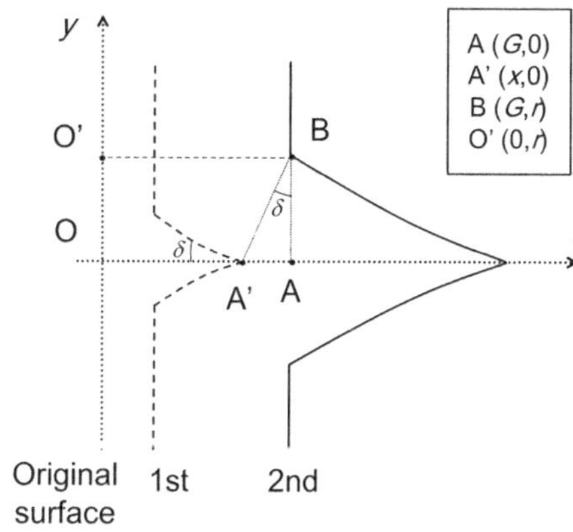

Fig. 7.4 Schematic cross-sectional view of the etch pit considered with the depth dependence of the track etching speed. The dotted line shows the original surface. The dashed and solid lines show the 1st and 2nd steps of the etch pit profiles

The sensitivity of an SSNTD should be considered with the depth dependence of the track etching speed. Figure 7.4 shows the etch pit profile with the depth dependence of the track etching speed. The geometry of each point of the etch pit is named as shown in Fig. 7.4. An ion entered the SSNTD material with normal incidence from point O. The horizontal axis x indicates the depth in the detector, and the vertical axis y indicates the radial direction. The dotted line shows the original surface before etching. The dashed and solid lines show the 1st and 2nd steps of the etch pit profiles, respectively. Point B was created by the route $O'B$, occupied only by the bulk etching process, and routes OA' (track etching) and $A'B$ (bulk etching) with the same etching time. Thus, (7.4) can be obtained as follows,

$$\frac{\overline{O'B}}{V_b} = \int_0^x \frac{d_x}{V_t(x)} + \frac{\overline{A'B}}{V_b}, \tag{7.4}$$

where $V_t(x)$ is the track etching speed as a function of depth in the detector. The left side of (7.4) indicates the time to etch $\overline{O'B}$ according to the bulk etching speed, and the right side indicates the time to etch $\overline{O'A}$ with the track etching speed and $\overline{A'B}$ with the bulk etching speed. The distances of $\overline{A'A}$ and $\overline{A'B}$ can be described as follows using the coordinates of the point in Fig. 7.4 and (7.2).

$$\overline{AA'} = r\frac{1}{\sqrt{S(x)^2 - 1}} = G - x, \tag{7.5}$$

$$\overline{A'B} = \frac{G - x}{\sin\delta} = r\frac{S(x)}{\sqrt{S(x)^2 - 1}} \tag{7.6}$$

Equation (7.7) is obtained by substituting (7.6) with (7.4) and multiplying both side by V_b, where $h(x)$ integrates the reciprocal of the sensitivity $S(x)$ over the depth as described in (7.8).

$$G = h(x) + r\frac{S(x)}{\sqrt{S(x)^2 - 1}}, \tag{7.7}$$

$$h(x) = \int_0^x \frac{1}{S(x)} \mathrm{d}x \tag{7.8}$$

From (7.5) and (7.8), the etch pit radius r and the thickness of layer removed G are described as follows.

$$r = \{x - h(x)\}\sqrt{\frac{S(x) + 1}{S(x) - 1}}, \tag{7.9}$$

$$G = \frac{xS(x) - h(x)}{S(x) - 1} \tag{7.10}$$

Thus, point B, the edge of the etch pit open mouth, has been described. In addition, the etch pit profile can be determined by the sensitivity $S(x)$, which is obtained by experiments.

The etch pit growth curve is defined as the growth of the etch pit radius r as a function of the thickness of layer removed G. The slope of the etch pit growth curve β is expressed as follows.

$$\beta = \frac{\mathrm{d}r}{\mathrm{d}G} = \frac{\left(\frac{\mathrm{d}r}{\mathrm{d}x}\right)}{\left(\frac{\mathrm{d}G}{\mathrm{d}x}\right)} = \sqrt{\frac{S(x) + 1}{S(x) - 1}} \tag{7.11}$$

The relation between the range of an ion R_0 and the residual range R, which is the distance between the etch pit tip and the end of the range, can be described as (7.12).

$$R = R_0 - x, \tag{7.12}$$

where x is the distance between R_0 and R. Using this relationship, the sensitivity can be rewritten as a function of the residual range R as follows.

$$S(R) = S(x) = \frac{1 + \beta^2}{1 - \beta^2}, \tag{7.13}$$

$$R = R_0 - \left(G - r\frac{1 - \beta^2}{2\beta}\right) \tag{7.14}$$

Based on these equations, the sensitivity of an SSNTD can be calculated using the parameter obtained by microscopic observations.

7.2.3 Determination of the Range of an Ion in an SSNTD

The range of an ion can be determined by the growth behaviour in the spherical phase of the etch pit, so-called rounded out, in which the etching exceeds the range of an ion. Figure 7.5 shows the etch pit profiles in the conical and spherical phases, i.e., before and after being rounded out, with dashed and solid lines, respectively. The depth of the etch pit l never changes before and after being rounded out. The lengths of the triangle ABC are described as follows:

$$\overline{AB} = G - R_0, \tag{7.15}$$

$$\overline{BC} = r, \tag{7.16}$$

$$\overline{CA} = G - (R_0 - l) \tag{7.17}$$

Applying the Pythagorean theorem to the triangle ABC, r^2 and R_0 can be derived as follows:

$$r^2 = 2lG - 2R_0 l + l^2, \tag{7.18}$$

$$R_0 = G + \frac{1}{4}\left(\frac{\mathrm{d}r^2}{\mathrm{d}G}\right) - \frac{r^2}{\left(\frac{\mathrm{d}r^2}{\mathrm{d}G}\right)} \tag{7.19}$$

From (7.18), therefore, the growth curve of the squared radius in the spherical phase is linear. The range R_0 can be determined from the slope of the linear function, i.e., $\mathrm{d}r^2/\mathrm{d}G = 2l$, using (7.19).

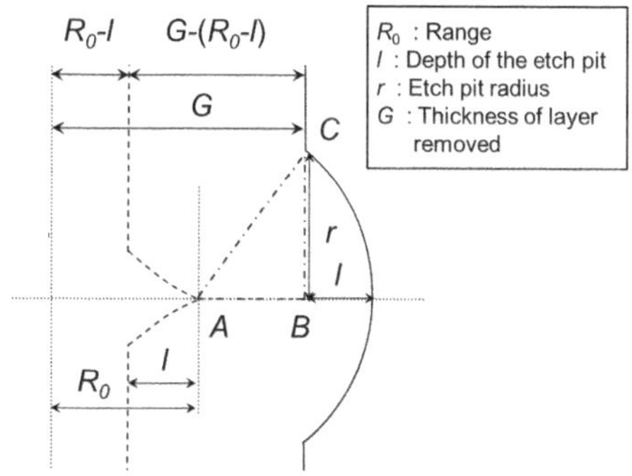

Fig. 7.5 Schematic cross-sectional view of the etch pit profiles in the conical and spherical phases. The dotted line shows the original surface. The dashed and solid lines show the conical and spherical etch pit profiles, respectively

7.2.4 Proof-of-Principal Experiments of Range Determination Using CR-39

α-particles with energies of 1.68, 2.77, 3.83, 4.77 MeV from an ^{241}Am source irradiated CR-39 detectors (BARYOTRAK, Fukuvi Chemical Industry). The energies were controlled by the distance in air using a variable-thickness collimator. The irradiated sample was chemically etched in a stirred 6 M KOH solution kept at 70 °C. To obtain the growth curves of the etch pits at each energy, chemical etching and microscope observations were repeated every 30 min, which is referred to as a multi-step etching technique.

Note that the radii of the etch pits were measured by reflected-light optical microscopy. The accuracy of the measurements, which depends on the resolution of the optical microscope, is approximately 0.1 μm. On the other hand, the thickness of layer removed usually includes an error of ±0.5 μm when using a conventional micrometre. In addition, the thickness of CR-39 detectors after chemical etching is difficult to measure by a micrometre. This is because CR-39 detectors swell by water absorption during chemical etching. To reduce such errors, we considered the average radius of the etch pits by fission fragments from a ^{252}Cf source as the thickness of layer removed. Figure 7.6 shows a schematic cross-sectional view of the etch pit by a fission fragment. A fission fragment, i.e., a heavy ion with low energy deposit a large amount of energy in the CR-39 detector with a range of less than a few tens of μm; that is, a dense damaged region is created in CR-39. In such cases, the etching along the fission fragment track has a very short duration of chemical etching, because the

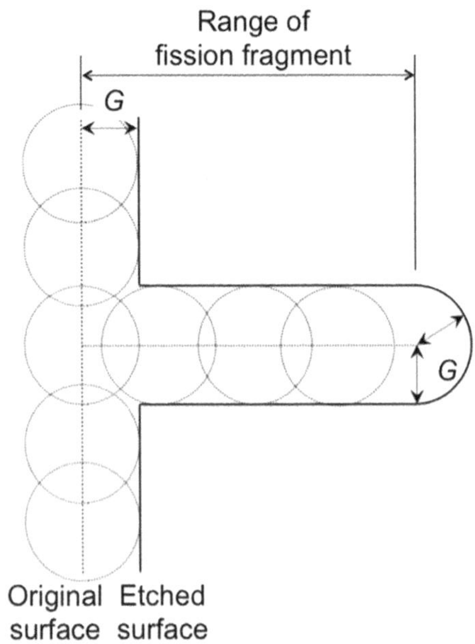

Fig. 7.6 Schematic cross sectional view of the etch pit by fission fragment. The radius corresponds to the thickness of laser removed, *G*

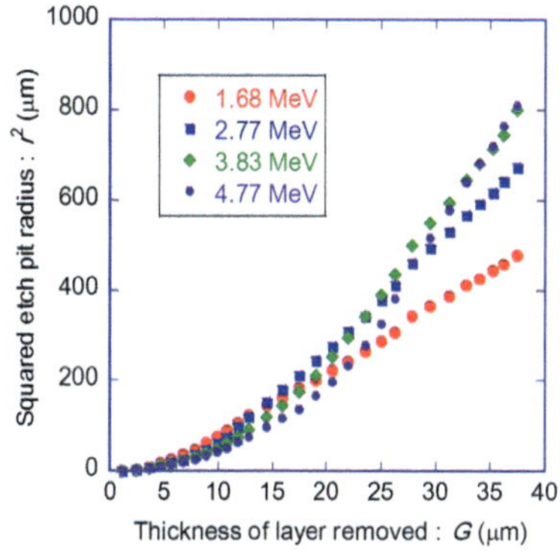

Fig. 7.7 Etch pit growth curve with multi-step etching, plotting the squared radius of the etch pits created by the α-particles with energies of 1.68, 2.77, 3.83, and 4.77 MeV

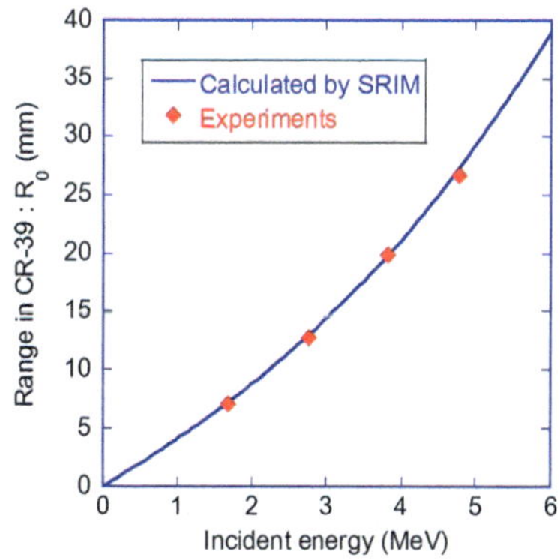

Fig. 7.8 Comparison between the ranges obtained in CR-39 detectors by the etch pit growth behaviour and the calculated ranges by the SRIM code

V_t is more than a few hundred times faster than V_b. In other words, the etch pit wall is formed practically in accordance with V_b. Therefore, the radius of the etch pit by the fission fragment corresponds to the thickness of layer removed G, as shown in Fig. 7.6. Using this principle and optical microscopy, the thickness of layer removed can be evaluated with almost the same accuracy as that of the etch pit radius.

Figure 7.7 shows growth curves of the squared etch pit radius as a function of the thickness of layer removed. Substituting the etch pit radii r, the thicknesses of layer removed G, and the slope of the straight part of the growth curves for (7.19), the ranges at each energy are determined. Figure 7.8 shows the obtained ranges in CR-39 detectors for the etch pit growth behaviour compared with the calculated values by the SRIM code [24]. The evaluated ranges are in good agreement with those of the calculations. This analysis can be applied to the evaluation of the energy spectrum of laser-accelerated ions as described in Sect. 7.3.

7.3 Application of CR-39 Detectors to Measurements of Laser-Accelerated Ions

Utilizing the explained analysis methods for etch pits in CR-39 detectors, this section introduces the application of CR-39 detectors to measurements of laser-accelerated ions from a cluster-gas target based on a previously published paper by the authors' group [25].

7.3.1 Experiments

Laser-driven ion acceleration experiments using a cluster-gas target have been conducted with the J-KAREN laser at the Kansai Photon Science Institute (KPSI). J-KAREN is an intense Ti:sapphire laser system that utilizes the optical parametric chirped pulse amplification (OPCPA) technique [26]. In the present experiment, 40 fs pulses with a pulse energy of 1 J were delivered to the experimental chamber. The laser pulses were focused with an off-axis parabolic mirror with focal length of 475 mm, leading to a peak intensity of 1×10^{19} W/cm^2 in vacuum. The cluster-gas target was produced by expanding a 6 MPa mixture of 70% H_2 and 30% CO_2 gas into vacuum using a pulsed valve connected with a special nozzle consisting of three truncated cones with different apex angles [27]. The nozzle has the capability to produce CO_2 clusters with an average diameter of 250 nm [28]. The laser beam was focused at the centre and entrance of the gas jet, 1.5 mm above the nozzle orifice.

To measure the candidate ions, i.e., protons from the background gas and carbon/oxygen ions from the clusters, two CR-39 detectors (HAZLAS (TD-1), Fukuvi Chemical Industry) covered by a 6 μm Al filter were installed behind the permanent magnet as shown in Fig. 7.9. The experimentally measured magnetic field was reconstructed in the three-dimensional electromagnetic field analysis software AMaze (Field Precision LLC), and the trajectories of protons with energies of 0.55–2.5 MeV and fully striped carbon ions with energies of 0.6–2.5 MeV/n from the focus point of the laser were calculated to decide the positions of the CR-39 detectors. Note that fully striped oxygen ions with the same kinetic energy should follow the same trajectories as the carbon ions because the charge to mass ratio is the same

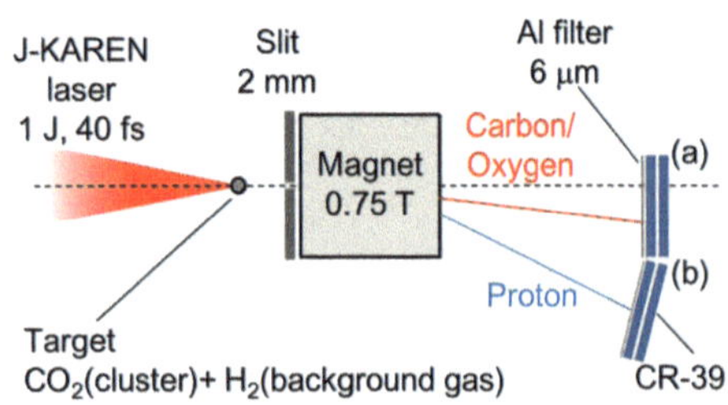

Fig. 7.9 Schematic top view of the experimental configuration. Carbon/oxygen ions and protons are deflected by the magnetic field and enter the CR-39 detectors (**a**) and (**b**), respectively

as that of the fully striped carbon ions. Additionally, carbon/oxygen ions with an energy lower than 0.6 MeV/n cannot penetrate the Al filter with a thickness of 6 μm. To measure carbon/oxygen ions, the detector (a) was installed at the position for the carbon/oxygen ions with energies larger than 0.6 MeV/n. On the other hand, the detector (b) was installed at the position for the carbon/oxygen ions with energies smaller than 0.6 MeV/n to measure the protons. In addition, the width of the slit before the magnet was also determined to be 2 mm to avoid overlapping protons and carbon/oxygen ions. Based on the calculation results, the CR-39 detectors (a) and (b) can simultaneously measure carbon/oxygen ions with energies larger than 0.7 MeV/n and protons with an energy range of 0.55–2.5 MeV, respectively.

After accumulating 1000 laser shots, CR-39 samples were chemically etched in a stirred 6 M KOH solution kept at 70 °C using the multi-step etching technique to obtain the etch pit growth curve. To obtain a spatial distribution of the etch pits on the etched samples, 1×6.5 cm^2 surfaces of the CR-39 samples were scanned using a fast automated digital imaging optical microscope (HSP-1000, Seiko Precision Inc.). The structure of the etch pits open mouths was processed with the HspFit software (Seiko Precision Inc.).

7.3.2 Energy Spectra of Laser-Accelerated Ions

Spatial distributions of the etch pits with etching durations of 2 h for the CR-39 detector (a) and 5 h for the CR-39 detector (b) are shown in Fig. 7.10. Each red dot corresponds to an etch pit. Note that the false etch pits, i.e., noise, on the order of 10–50 pits/cm^2, which are created by the scattered ions and surface roughness, are not included in Fig. 7.10. The lower energy particles were deflected by the magnetic field and entered the right side of CR-39 detectors. Typical images of the etch pits on the CR-39 detectors, obtained by optical microscopy, are also shown in Fig. 7.10. Larger etch pits are observed on CR-39 detector (a) than on CR-39 detector (b), which indicates that the heavier ions, i.e., carbon/oxygen ions, were incident on the

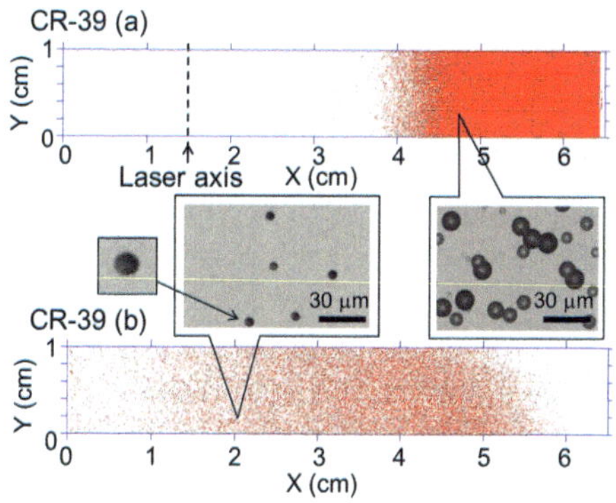

Fig. 7.10 Spatial distributions of the etch pits in the CR-39 detectors. Each red dot corresponds to an etch pit. The magnified views by optical microscopy show the etch pits of carbon/oxygen in CR-39 detector (**a**) and etch pits of protons with oblique incidence in CR-39 detector (**b**)

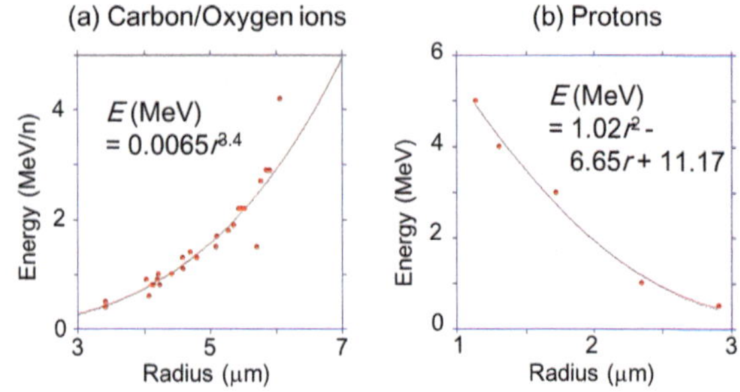

Fig. 7.11 Calibration curves for carbon/oxygen ion obtained by the multi-step etching technique (**a**) and for protons with the energy of 0.5–5 MeV by conventional accelerator (**b**)

CR-39 detector (a). Note that the etch pits on CR-39 detector (b) show an elliptical open mouth, as shown in the magnified view of Fig. 7.10. These etch pits indicated that the incidence angle was not perpendicular to the surface. Because protons were deflected by the magnetic field and obliquely entered the CR-39 detector (b). The etch pits on CR-39 detector (a) with a brighter centre part indicate that the etch pit tips were already in the spherical phase after 2 h of etching. On the other hand, etch pits of protons on the CR-39 detector (b) were still in the conical phase after 5 h of etching. This is because the projected range of a few MeV protons in the CR-39 detector is longer than that of carbon/oxygen ions of a few MeV/n. Therefore, different analysis methods should be applied to obtain the energy of carbon/oxygen ions and protons as follows.

To evaluate the energies of carbon/oxygen ions, the ranges in CR-39 detector (a) are obtained by the range determination method described in Sects. 7.2.3 and 7.2.4. We randomly picked approximately 20 etch pits and applied the multi-step etching technique to CR-39 detector (a). From the etch pit growth curves of these etch pits, the ranges for each etch pit were obtained using (7.19). Finally, the incident energies for each etch pit were calculated based on the obtained ranges by the SRIM code. The calibration curve, i.e., correlation between the ranges in CR-39 detector (a) and the incident energies, for carbon/oxygen ions is shown in Fig. 7.11a. The higher energy particle creates a larger-sized etch pit in the spherical phase, because the higher energy particle range in the CR-39 detector is deeper than the lower energy particle range. On the other hand, the energies of protons in CR-39 (b) were evaluated by the calibration curve obtained by using conventional accelerator, as shown in Fig. 7.11b. This calibration curve indicates that the higher energy proton creates a smaller-sized etch pit.

Based on these calibration curves, the energy of each etch pit in CR-39 detectors (a) and (b) and the energy spectra were evaluated as shown in Fig. 7.12. The maximum energies of the carbon/oxygen ions and protons were 1.1 ± 0.1 MeV/n and 1.6 ± 0.1 MeV, respectively. In the case of carbon/oxygen ions, the number of ions sharply decreases in the higher energy region. From the energy of the carbon/oxygen ions, it can be concluded that carbon/oxygen ions are accelerated by hydrodynamic ambipolar expansion, where only some of the electrons are blown off from the cluster;

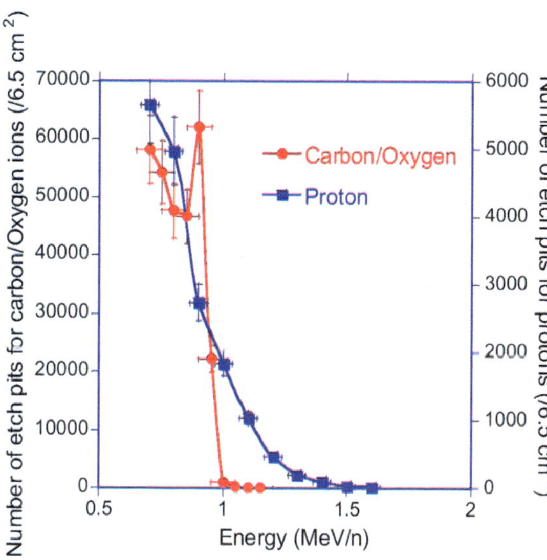

Fig. 7.12 Energy spectra of laser-accelerated carbon/oxygen ions and protons

i.e., some electrons remain inside of the cluster [29, 30]. On the other hand, the number of protons gradually decreases in the higher energy region, this trend is different from that of the carbon/oxygen ions, which could be related to the dynamics of the boundary layer in between two distinct collisionless plasmas [31].

7.4 Conclusions

The fundamentals of SSNTD were described based on the studies of pioneering researchers. Additionally, the method of using CR-39 detectors as an ion detector in laser-driven ion acceleration experiments was explained using concrete examples. In particular, the multi-step etching technique provided precise energy spectra of laser-accelerated ions with an energy resolution of $\Delta E = 0.1$ MeV. Such precise energy spectra may allow us to discuss the ion acceleration mechanism in terms of the interaction between target materials and intense laser pulses. Therefore, CR-39 detectors are the best candidate detectors for measurements of laser-accelerated ions.

Acknowledgements This work was partially supported by the Funding Program for Next Generation World-Leading Researchers (NEXT Program), Grant-in-Aids for Scientific Research No. 26247100, No. 19H00668 and No. 17K17876 by JSPS KAKENHI and the Consortium for Photon Science and Technology (CPhoST) program by MEXT.

References

1. A. Macchi, M. Borghesi, M. Passoni, Ion acceleration by superintense laser-plasma interaction. Rev. Mod. Phys. **85**, 751–793 (2013)
2. H. Daido, M. Nishiuchi, A.S. Pirozhkov, Review of laser-driven ion sources and their applications. Rep. Prog. Phys. **75**, 056401 (2012)
3. A. Higginson, R.J. Gray, M. King, R.J. Dance, S.D.R. Williamson, N.M.H. Butler, R. Wilson, R. Capdessus, C. Armstrong, J.S. Green, S.J. Hawkes, P. Martin, W.Q. Wei, S.R. Mirfayzi, X.H. Yuan, S. Kar, M. Borghesi, R.J. Clarke, D. Neely, P. McKenna, Near-100 MeV protons via a laser-driven transparency-enhanced hybrid acceleration scheme. Nat. Comm. **9**, 724 (2018)
4. S. Gaillard, J. Fuchs, N. Renard-Le Galloudec, T.E. Cowan, Study of saturation of CR39 nuclear track detectors at high ion fluence and of associated artifact patterns. Rev. Sci. Instrum. **78**, 013304 (2007)
5. Y. Fukuda, A.Ya. Faenov, M. Tampo, T.A. Pikuz, T. Nakamura, M. Kando, Y. Hayashi, A. Yogo, H. Sakaki, T. Kameshima, A.S. Pirozhkov, K. Ogura, M. Mori, T.Zh. Esirkepov, J. Koga, A.S. Boldarev, V.A. Gasilov, A.I. Magunov, T. Yamauchi, R. Kodama, P.R. Bolton, Y. Kato, T. Tajima, H. Daido, S.V. Bulanov, Energy increase in multi-MeV ion acceleration in the interaction of a short pulse laser with a cluster-gas target. Phys. Rev. Lett. **103**, 165002 (2009)
6. D. Haberberger, S. Tochitsky, F. Fiuza, C. Gong, R.A. Fonseca, L.O. Silva, W.B. Mori, C. Joshi, Collisionless shocks in laser-produced plasma generate monoenergetic high-energy proton beams. Nat. Phys. **8**, 95–99 (2012)
7. T. Yamauchi, H. Ichijo, K. Oda, Depth-dependence of the bulk etch rate of gamma-ray irradiated CR-39 track detector. Radiat. Meas. **28**, 191–196 (2001)
8. G. Somogyi, S.A. Szalay, Track-diameter kinetics in dielectric track detectors. Nucl. Instrum. Meth. **109**, 211–232 (1973)
9. K. Oda, I. Csige, R.P. Henke, E.V. Benton, A new method for internal calibration of nuclear track detectors. Int. J. Radiat. Appl. Instrum. Nucl. Tracks Radiat. Meas. **20**, 505–510 (1992)
10. T. Yamauchi, K. Oda, S. Tanabe, H. Matsumoto, H. Miyake, Calibration study on the triton response of CR-39 track detector. Radiat. Meas. **28**, 191–196 (1997)
11. M. Kanasaki, A. Hattori, H. Sakaki, Y. Fukuda, A. Yogo, S. Jinno, M. Nishiuchi, K. Ogura, K. Kondo, K. Oda, T. Yamauchi, A high energy component of the intense laser-accelerated proton beams detected by stacked CR-39. Radiat. Meas. **50**, 46–49 (2013)
12. H. Sakaki, M. Kanasaki, Y. Fukuda, M. Nishiuchi, T. Hori, A. Yogo, S. Jinno, K. Niita, Development of a single-shot-imaging thin film for an online Thomson parabola spectrometer. Rev. Sci. Instrum. **84**, 013301 (2013)
13. T.W. Jeong, P.K. Singh, C. Scullion, H. Ahmed, P. Hadjisolomou, C. Jeon, H. Yun, K.F. Kakolee, M. Borghesi, S. Ter-Avetisyan, CR-39 track detector for multi-MeV ion spectroscopy. Sci. Rep. **7**, 2152 (2017)
14. Y. Mori, T. Yamauchi, M. Kanasaki, Y. Maeda, K. Oda, S. Kodaira, T. Konishi, N. Yasuda, R. Barillon, Radiation chemical yields for loss of ether and carbonate ester bonds in PADC films exposed to proton and heavy ion beams. Radiat. Meas. **46**, 1147–1153 (2011)
15. B.G. Cartwright, E.K. Shirk, P.B. Price, A nuclear-track-recording polymer of unique sensitivity and resolution. Nucl. Instrum. Methods. **153**, 457–460 (1978)
16. J. Drach, P.B. Price, M.H. Salamon, Characteristics of cronar polyethylene terephthalate track detectors. Nucl. Instrum. Methods. Phys. Res. Sec. B. **28**, 49–52 (1987)
17. R.L. Fleischer, P.B. Price, R.M. Walker, Criterion for registration in dielectric track detectors. Phys. Rev. **156**, 353–355 (1967)
18. R. Barillon, M. Fromm, A. Chambaudet, H. Marah, A. Sabir, Track etch velocity study in a radon detector (LR 115, cellulose nitrate). Radiat. Meas. **28**, 619–628 (1997)
19. K. Ogura, T. Hattori, M. Hirata, M. Asano, M. Yoshida, M. Tamada, H. Omichi, N. Nagaoka, H. Kubota, R. Katakai, Development of copolymer of CR-39 with high sensitivity to low let particles. Radiat. Meas. **25**, 159–162 (1995)

20. K. Ogura, M. Hattori, M. Asano, M. Yoshida, H. Omichi, N. Nagaoka, H. Kubota, R. Katakai, H. Hasegawa, Proton response of high sensitivity CR-39 copolymer. Radiat. Meas. **28**, 197–200 (1997)
21. K. Ogura, M. Asano, N. Yasuda, M. Yoshida, Properties of TNF-1 track etch detector. Nucl. Instrum. Methods. Phys. Res. Sec. B. **185**, 222–227 (2001)
22. R.P. Henke, E.V. Benton, On geometry of tracks in dielectric nuclear track detectors. Nucl. Instrum. Methods. **97**, 483–489 (1971)
23. G. Somogyi, Development of etched nuclear tracks. Nucl. Instrum. Methods. **173**, 21–42 (1980)
24. J.F. Ziegler, M.D. Ziegler, J.P. Biersack, SRIM—the stopping and range of ions in matter. Nucl. Instrum. Methods. Phys. Res. Sec. B. **268**, 1818–1823 (2010)
25. M. Kanasaki, S. Jinno, H. Sakaki, K. Kondo, K. Oda, T. Yamauchi, Y. Fukuda, The precise energy spectra measurement of laser-accelerated MeV/n-class high-Z ions and protons using CR-39 detectors. Plasma Phys. Control. Fusion. **58**, 034013 (2016)
26. H. Kiriyama, M. Mori, Y. Nakai, T. Shimomura, H. Sasao, M. Tanaka, Y. Ochi, M. Tanoue, H. Okada, S. Kondo, S. Kanazawa, A. Sagisaka, I. Daito, D. Wakai, F. Sasao, M. Suzuki, H. Kotakai, K. Kondo, A. Sugiyama, S. Bulanov, P.R. Bolton, H. Daido, S. Kawanishi, J.L. Collier, C. Hernandez-Gomez, C.J. Hooker, K. Ertel, T. Kimura, T. Tajima, High-spatiotemporal-quality petawatt-class laser system. Appl. Opt. **49**, 2105–2115 (2010)
27. S.V. Bulanov, D.V. Dylov, T.Zh. Esirkepov, F.F. Kamenets, D.V. Sokolov, Ion acceleration in a dipole vortex in a laser plasma corona. Plasma Phys. Rep. **31**, 409–421 (2006)
28. S. Jinno, Y. Fukuda, H. Sakaki, A. Yogo, M. Kanasaki, K. Kondo, A.Ya. Faenov, I.Yu. Skobelev, T.A. Pikuz, A.S. Boldarev, V.A. Gasilov, Mie scattering from submicron-sized CO_2 clusters formed in a supersonic expansion of a gas mixture. Opt. Express. **21**, 20656–20674 (2013)
29. Y. Kishimoto, T. Masaki, T. Tajima, High energy ions and nuclear fusion in laser-cluster interaction. Phys. Plasmas. **9**, 589–601 (2002)
30. M. Kanasaki, S. Jinno, H. Sakaki, A.Ya. Faenov, T.A. Pikuz, M. Nishiuchi, H. Kiriyama, M. Kando, A. Sugiyama, K. Kondo, R. Matsui, Y. Kishimoto, K. Morishima, Y. Watanabe, C. Scullion, A.G. Smyth, A. Alejo, D. Doria, S. Kar, M. Borghesi, K. Oda, T. Yamauchi, Y. Fukuda, Observation of the inhomogeneous spatial distribution of MeV ions accelerated by the hydrodynamic ambipolar expansion of clusters. Radiat. Meas. **83**, 12–14 (2015)
31. R. Matsui, Y. Fukuda, Y. Kishimoto, Dynamics of the boundary layer created by the explosion of a dense object in an ambient dilute gas triggered by a high power laser. Phys. Rev. E **100**, 013203 (2019)

Chapter 8
Towards Laser Intensity Calibration Using High-Field Ionization

M. F. Ciappina, S. V. Bulanov, T. Ditmire, G. Korn, and S. Weber

Abstract We present an approach for direct measurement of ultrahigh laser intensities in the range 10^{20}–10^{24} W/cm^2. The method is based on the use of multiple sequential tunneling ionization of heavy atoms with adequately high ionization potentials. We show that, due to the highly nonlinear dependence of tunneling ionization rates on the electromagnetic field strength, an off-set in the charge distribution of ions appears to be clearly sensitive to the peak value of intensity in the laser focus. Based on the tunnel-ionization mechanism, a simple analytic theory helps in estimating the maximal charge state produced at a given laser intensity. Our theory also allows for calculating qualitatively a distribution in charge states generated in different zones of the laser focus. These qualitative predictions are in excellent agreement with numerical simulations of the tunneling cascades, developed in the interaction of a short tightly focused laser pulse with low-density noble gas targets. The method could be particularly useful and of instrumental demand in view of the expected commissioning of several new laser facilities, capable of delivering ultra-powerful light pulses in the above mentioned domain of intensities.

8.1 Introduction

The continuous development during the last two-decades of high-power laser sources operating at optical and infrared wavelengths, has lead to a noticeable growth in intensities of electromagnetic radiation available in laboratories. Present-day intensities

M. F. Ciappina (✉) · S. V. Bulanov · G. Korn · S. Weber
Institute of Physics of the ASCR, ELI-Beamlines project, Na Slovance 2,
182 21 Prague, Czech Republic
e-mail: marcelo.ciappina@eli-beams.eu

T. Ditmire
Department of Physics, University of Texas at Austin, Center for High Energy Density Science,
C1510, Austin, TX 78712, USA

S. Weber
School of Science, Xi'an Jiaotong University, Xi'an 710049, China

K. Yamanouchi and D. Charalambidis (eds.), *Progress in Ultrafast Intense Laser Science XV*, Topics in Applied Physics 136,
https://doi.org/10.1007/978-3-030-47098-2_8

approaching to 10^{19} W/cm^2 are being routinely used in many laser facilities. Furthermore, several lasers of petawatt (PW) power [1] could deliver pulses of intensities on the order of 10^{22} W/cm^2, depending on the size of the focal region. Singular reports of even higher intensities $\gtrsim 10^{22}$ W/cm^2 [2–4] are still lacking independent and unambiguous evidences. However, the forthcoming commissioning of several new 10-PW class laser facilities [5–10] opens a way to a new considerable step forward. For a laser power of 3 PW, expected to be reached at the ELI-Beamlines facility within the next 2–3 years, the peak intensity in a 3λ focal spot of an 800-nm laser pulse will exceed 5×10^{22} W/cm^2. With 10 PW laser pulses focused down to the diffraction limit, intensities on the order of $\simeq 10^{24}$ W/cm^2 or even higher can be achieved.

This experimental breakthrough is expected to make novel regimes of laser-matter interactions accessible for research including: the radiation-dominated regime, where radiation friction forces play the major role in plasma dynamics; observation and quantification of relativistic tunneling; generation of QED cascades of elementary particles developed from seed particles in a laser focus and a large zoo of other effects inaccessible at presently available laser intensities. For the overview of this newly emerging research area we direct the reader to reviews [11–13] and the literature quoted there.

In view of these prospects, the problem of precise and univocal determination of the electromagnetic field intensity in a laser focus becomes of singular importance. Here we propose using tunneling ionization of multicharged ions as a *probe* of laser intensity. The main advantage of this approach has its root in the highly nonlinear dependence of the tunneling ionization probability on the electromagnetic field strength, so that a relatively small change in intensity leads to orders of magnitude variation in the ionization rate. As a consequence, the maximal ionic charge observed in a laser focus becomes a highly sensitive function of the peak intensity. The shape of photoelectron spectra and positions of the spectral maximum in a circularly polarized field [14] and the high-energy cut-off in a linearly polarized field [15] are also sensitive to the value of intensity and have been extensively employed for its determination. In general, at moderate intensities $\mathcal{I} \leq 10^{15}$ W/cm^2, when single-electron ionization remains the dominant process, photoelectron spectra are the most efficient and reliable tool to measure the laser intensity [16, 17]. With its value growing, multiple ionization comes into play along with two competing mechanisms known as sequential and nonsequential ionization, with the latter triggered by the so-called *recollisions* [18–21]. Recollision takes place when the laser-ionized electron returns close to the ion core, driven by the laser electric field, that has now reversed its direction. In this re-encounter, this recolliding electron can elastically rescatter with the ion or trigger the above two mentioned processes. Interplay of the two mechanisms makes the interval $\mathcal{I} \simeq 10^{15}$–$10^{17}$ W/cm^2 difficult for unambiguous determination of intensity via the ionization process. A further increase of intensity simplifies the situation again. As soon as the electron motion after the ionization step becomes relativistic which, for $\lambda = 800$ nm happens at $\mathcal{I} \approx 10^{17}$ W/cm^2, the longitudinal photoelectron drift induced by the Lorentz force makes the recollision with the ion core impossible, so that the sequential tunnel ionization remains the only relevant and dominant mechanism. Simultaneously, deeper electronic shells are stripped out by the laser: helium is fully ionized at $\mathcal{I} \approx 5 \times 10^{16}$ W/cm^2 and neon at $\mathcal{I} \approx 10^{20}$ W/cm^2. Thus probing higher intensities with ionization requires a target consisting of high-Z atoms [22, 23].

A substantial number of experimental works considered multiple ionization of atoms by intense optical and infrared laser radiation. As far as ultrahigh intensities are concerned, Walker et al. [24, 25] and Yamakawa et al. [26, 27] considered ionization of noble gases at intensities up to 10^{19} W/cm^2, aiming to validate the tunneling ionization theory, assuming the value of intensity known with certain degree of accuracy. Although in their calculations they did not consider effects of the intensity distribution over the laser focus and simply estimated the maximal ionic charge number using the tunneling rate formulas [28–32], a fairly good agreement with the experimental data was demonstrated. Particularly for the case of [26, 27] the utilization of a mobile slit to limit the intensity-dependent volume expansion, placed on the central axis of the time of flight (TOF) tube (see e.g. [33] for details), made possible to detect ions only coming from the tight-focused region (the most intense one) and in this way improving the comparison with the theoretical models. In other high field experiments within the tunneling regime, the ionization yield of He was employed to confirm focusing intensities of over 10^{20} W/cm^2 with an excellent precision, less than 20%, when compared with optical measurements of the laser pulse [33].

In this contribution, we suggest using a setup similar to the one implemented in [24–27] in the opposite way: the unknown value of the peak intensity and, to some extent, of the spatial intensity distribution in the focus can be reconstructed, in principle, from the distributions of ionic charge states. This idea has been rather widely spread in the strong field community (see, e.g. [22, 23]) with, however, little emphasis on both the experimental constraints needed for its realization and to the accuracy which can be achieved within this method. Additionally, to the best of our knowledge, a systematic and exhaustive theoretical study of this approach, joint with its limitations and advantages, is still missing.

In order to test the feasibility of this idea, we employ the theory of sequential tunnel ionization and numerically obtain the charge spectra by varying the laser peak value. Our article is organized in the following way. Section 8.2 presents qualitative estimates justifying the theory of single-particle nonrelativistic sequential tunneling, which applies for the problem under study up to intensities of $\mathcal{I} \approx 10^{24}$ W/cm^2. Here, a brief summary of the tunneling ionization theory is given as well, in the part which relates to the total ionization rates. Using an analytic formulae for the rate of tunneling ionization, a simple estimate for the off-set ionization potential, corresponding to the highest charge state present in the spectrum, is presented. In Sect. 8.3 we formulate the system of rate equations, based on the ionization rates, that govern the laser ionization of argon, krypton and xenon. Essential simplifications of the full system, making it accessible for fast and easy to handle numerical computations, are introduced and discussed here. Results of numerical calculations are presented as well and compared to the theoretical predictions of Sect. 8.2. Finally, the last section contains a brief conclusion. As a matter of completeness, we include in the Appendices a complete description of the systems of rate equations, as well as all the parameters needed for their numerical implementation. Atomic units $e = m_e = \hbar = 1$, with the speed of light $c = 1/\alpha$, where α is the fine structure constant, are used unless stated otherwise.

8.2 Theory

8.2.1 Qualitative Analysis

We rely on the theoretical approach introduced in [34]. It roots on the following physical considerations:

1. A gas in the laser focus is kept at sufficiently low pressure, so that neither nonlinear propagation and plasma effects in the medium, nor electron-ion collisions, play any relevant role. For instance, at a concentration of atoms $n_0 = 10^{14}$ cm^{-3} and an average ionic charge $\bar{z} = 20$, the plasma frequency is of the order of $\omega_p \simeq 10^{12}$ s^{-1}, so that no plasma oscillations take place during the interaction with a laser pulse of 100-fs or shorter time duration. The energies of the laser-ionized photoelectrons ε_e are of the order of the ponderomotive energy, so that in the ultrarelativistic case we are interested in results:

$$\varepsilon_e \simeq \sqrt{1 + \frac{a_0^2(1+\rho^2)}{2}} c^2 \simeq a_0 c^2, \tag{8.1}$$

where a_0 is the dimensionless field amplitude defined as:

$$a_0 = \frac{E_0}{\omega c}, \tag{8.2}$$

with E_0 and ω being the laser field amplitude and frequency, respectively. Equation (8.1) gives energies $\varepsilon_e \approx 4 \times 10^6$ eV and $\varepsilon_e \approx 4 \times 10^8$ eV for $\mathcal{I} = 10^{20}$ W/cm^2 and $\mathcal{I} = 10^{24}$ W/cm^2, respectively. In (8.1) ρ defines the laser field ellipticity, with $\rho = 0$ ($\rho = \pm 1$) for linear (elliptical) polarization. At such energies, cross-sections of elastic and inelastic electron ion collisions are comparable to that of bremsstrahlung, with none of them exceeding 10^{-20} cm^2, even at the lower boundary of the intensity interval (for numerical values and formulas see e.g. [35, 36]). As a result, the mean free path of electrons exceeds the laser focal size (which is limited by several microns for the tight focusing necessary to reach the highest values of intensity, see Sect. 2.4) by several orders in magnitude. At the same time, the number of atoms in the interaction volume, estimated assuming the diffraction-limit focusing $N_a \simeq n_0 \lambda^3 \approx 10^2$, remains sufficiently high to reliably record the distribution in charge states with a typical TOF ion detector.

2. We a priori assume multiple ionization to be sequential, so that electrons leave an atom independently, i.e. one after another. This assumption relies on two inequalities. First, the travel time of the electron under the potential barrier

$$\tau_{\rm sb} \simeq \frac{b}{v_0} \approx \frac{\sqrt{I_p}}{\sqrt{2}E_0}, \tag{8.3a}$$

where $b = I_p/E_0$ is the barrier width and $v_0 = \sqrt{2I_p}$ is the characteristic electron velocity in a bound state with ionization potential $I_{\rm p}$, is much longer than the electron-electron or electron-ion correlation time for the same bound state

$$\tau_{\rm corr} \simeq \frac{a_B}{v_0} \approx \frac{1}{2I_p}, \tag{8.3b}$$

i.e. $\tau_{\rm sb} \gg \tau_{\rm corr}$. In (8.3b) $a_B \simeq 1/\sqrt{2I_p}$ is the characteristic size of the bound state (Bohr radius). The ratio of these two times results then:

$$\frac{\tau_{\rm corr}}{\tau_{\rm sb}} \simeq \frac{E_0}{E_{\rm ch}} \equiv F, \quad E_{\rm ch} = (2I_p)^{3/2}, \tag{8.3c}$$

where $E_{\rm ch}$ is the characteristic electric field strength at the Bohr orbit of the bound state, and F is known as the reduced field [30]. For tunneling or multiphoton ionization the reduced field is always numerically small, i.e. $F \ll 1$. Furthermore, its typical value decreases with increase of both the atomic ionization potential and laser intensity, as we will demonstrate below.

3. Although the electron motion after the ionization step quickly becomes ultra-relativistic, the tunneling itself proceeds nonrelativistically, as long as the ionization potential is small compared to the rest energy of the electron, i.e. $I_p \ll c^2$. Quantitatively, relativistic effects in laser-induced tunneling are determined by the value of the parameter [30, 32]

$$\xi^2 = 1 - \frac{1}{2}\epsilon\left[\sqrt{\epsilon^2+8} - \epsilon\right], \quad \epsilon = \frac{c^2 - I_p}{c^2}. \tag{8.4a}$$

In all the cases we study here, ξ remains small, i.e.

$$\xi^2 \approx \frac{2I_p}{3c^2} \ll 1. \tag{8.4b}$$

As an example, let us consider the ground $1s$ state of the Ar^{17+} ion, with $I_p =$ 4426 eV $\approx$ 163 a.u. Below we show that intensities $\mathcal{I} \approx 2 \times 10^{21}$ W/cm^2 are required to ionize this state producing bare Ar^{18+} ions. In this case results $\xi^2 = 0.0058$, showing that the relativistic effect on the tunneling remains on the level of 1% or less. For intensities $\mathcal{I} \approx 10^{24}$W/cm^2 and $I_p \approx 30$ keV (see Fig. 8.1) $\xi^2 \approx 0.04$, so that the nonrelativistic approximation remains quite accurate even at such otherwise ultra relativistic intensities. This considerable extension of the nonrelativistic regime of tunneling toward ultra-high intensities, roughly six orders in magnitude above that corresponding to $a_0 = 1$, where the electron motion becomes fully relativistic, can be qualitatively explained in the following way. The ionization step can be viewed as a detachment of the electron from its atomic orbital. This event takes place at a distance of the order of the tunnel barrier width:

$$b = \frac{I_p}{E_0} \gg a_B. \tag{8.5}$$

The electron travels this distance b during the time given by (8.3a), which is a small fraction of the laser period: the respective time can be estimated as:

$$\tau_{\rm sb}\omega = \frac{\sqrt{I_p}\omega}{\sqrt{2}E_0} \equiv \frac{\gamma}{2} \ll 1. \tag{8.6}$$

Here we introduce the well-known Keldysh parameter γ [37]. During this time interval the electron cannot be accelerated to relativistic kinetic energies, so that the magnetic component of the Lorentz force remains a small correction, unless the electron has had a relativistic velocity already in the beginning of its sub-barrier motion. The latter is only possible for $I_p \simeq c^2$. As a result, at $\xi^2 \ll 1$ (see (8.4a) and (8.4b)) the process of the electron's escape from an atom (ion) proceeds within the nonrelativistically domain.

4. Finally, note that ionization takes place in a deep tunneling regime, when the Keldysh parameter (8.6) is small. For a laser with wavelength $\lambda \simeq 1\,\mu$m the value of (8.6) varies from $\gamma \simeq 10^{-2}$ for $\mathcal{I} = 10^{20}$ W/cm^2 and $I_p = 10^3$ eV to $\gamma < 10^{-3}$ for $\mathcal{I} = 10^{24}$ W/cm^2 and $I_p = 3 \times 10^4$ eV. This means that instantaneous static-field ionization rates are entirely applicable for the calculation of the ionization probability. For the following, it will be useful to relate the laser electric field amplitude to the value of intensity expressed in units of 10^{20} W/cm^2 as:

$$E_0 = \frac{53.4\sqrt{\mathcal{I}}}{\sqrt{1+\rho^2}}, \tag{8.7}$$

where ρ is the laser field ellipticity. Throughout this contribution we set $\rho = 0$, i.e. we particularly consider the linear polarized case.

8.2.2 Tunneling Ionization Rates Calculation

Under the assumptions formulated in the previous subsection, the ionization rate of an ionic level with a residual charge z, i.e. $z = 1$ for neutral atoms and $z = N$ if all the electrons are removed (ionization of an $\mathrm{A}^{(N-1)+}$ ion), with the effective principal quantum number [30, 31]

$$\nu = \frac{z}{\sqrt{2I_p}}, \tag{8.8}$$

and orbital and magnetic quantum numbers l and m, is given by the well-known Perelomov-Popov-Terentiev (PPT) formula [28, 30]:

$$w(\nu, l, m; t) = C_{\nu l}^2 B_{lm} I_p F^{1+|m|-2\nu}(t) \exp\left\{-\frac{2}{3F(t)}\right\}, \tag{8.9a}$$

with

$$C_{\nu l}^2 = \frac{2^{2\nu-2}}{\nu\Gamma(\nu+l+1)\Gamma(\nu-l)}, \tag{8.9b}$$

$$B_{lm} = \frac{(2l+1)(l+|m|)!}{2^{2|m|}|m|!(l-|m|)!}, \tag{8.9c}$$

and the time-dependent reduced field $F(t)$ defined as:

$$F(t) = \sqrt{\frac{\mathbf{E}_L^2(t)}{(2I_p)^3}}, \tag{8.10}$$

where $\mathbf{E}_L(t)$ is the time-dependent laser electric field with a peak amplitude E_0. The asymptotic coefficient $C_{\nu l}$ in (8.9b) is taken in the approximate form introduced by Hartree [38]. Once that the ionization rates are known, it is possible to write a system of rate equations determining populations $0 \leq c_z(t) \leq 1$ of different ionic states in the form:

$$\sum_z c_z(t) = 1, \tag{8.11a}$$

$$c_z(t) = \sum_{j,l} c_z(j, l; t), \tag{8.11b}$$

$$\frac{\mathrm{d}c_z(j, l; t)}{\mathrm{d}t} = \sum_{j',l'} c_{z-1}(j', l'; t)\frac{n_{j',l'}}{2l'+1}\sum_{m'} w(\nu', l', m'; t) -$$

$$- c_z(j, l; t)\frac{n_{j,l}}{2l+1}\sum_{m''} w(\nu, l, m''; t). \tag{8.11c}$$

Here the index j denotes the ionization pathway (see the Appendices for more details and examples), $c_z(l, m; t)$ name the partial populations of orbitals with quantum number l along a fixed ionization pathway, and $n_{j,l}$ represents the number of equivalent electrons at the orbital. The system of rate equations is truncated either at $z = Z$, where the Z is the atomic number of the element, or earlier if the intensity is not enough to fully ionize the atom. It can also be truncated from the side of small z, owing to the fact that all states with relatively small ionization potentials appear to be quickly depleted at high intensities. Abbreviated systems of rate equations for argon, krypton, and xenon are presented in the Appendices.

8.2.3 Intensity-Dependent Ionization Offset

Before discussing the numerical solution of the system (8.11a)–(8.11c), which is done in the next section, we obtain an estimate for the maximal charge state which can be achieved via tunneling ionization for a given value of laser intensity. The purpose of this approximate calculation is twofold. First, it allows for approximately finding, without stringent numerical calculations, the interval of charge states expected to be observed if the value of intensity is known, and vice versa. Secondly, and not less relevant, it will help us to considerably reduce the number of terms in the system of equations defined in (8.11a)–(8.11c) by omitting those ones which, at a given intensity, correspond either to levels which are ionized instantaneously, so that for them $c_z = 0$ or, instead, to those which stay almost unaffected for the laser field, i.e. $c_z \approx 1$. For high-Z ions, where the number of rate equations becomes excessively large, this markedly simplifies the actual numerical calculations.

In order to estimate the maximal charge number which can be produced at a given intensity with a high probability, we neglect the (l, m) dependence in (8.9a) and consider the total ionization probability per laser period for an s state at the maximal intensity (to simplify notations, we set the time such that the maximum is achieved at $t = 0$):

$$W = \frac{2\pi}{\omega} w(\nu, 0, 0; 0) = \pi C_\kappa^2 K_0 2^{2\nu+2} F^{1-2\nu} \exp\left\{ -\frac{2}{3F} \right\}, \tag{8.12}$$

where $K_0 = I_p/\omega$ is the multiquantum parameter [30, 31] and the reduced field F, see (8.10), is calculated for the peak amplitude value E_0 of the laser electric field. The exponential factor dominates the intensity dependence of the rate (8.12), while the prefactor plays a very minor role in the estimation of the ionization offset value. For a given value of intensity (and therefore of E_0) we estimate the threshold ionization potential $I_p^*(\mathcal{I})$, such that the probability of ionization per laser cycle is of the order of unity, $W \simeq 1$. This condition is essentially approximate as it ignores the pulse duration, the (l, m) dependence and the ionization saturation effect. However, thanks to the exponential factor in (8.12), which changes very rapidly with variations of the laser field amplitude, it provides a logarithmically accurate estimate:

$$F^* = \frac{2}{3} \ln^{-1} \left[2^{2\nu+2} C_\kappa^2 \pi K_0 (F^*)^{1-2\nu} \right]. \tag{8.13}$$

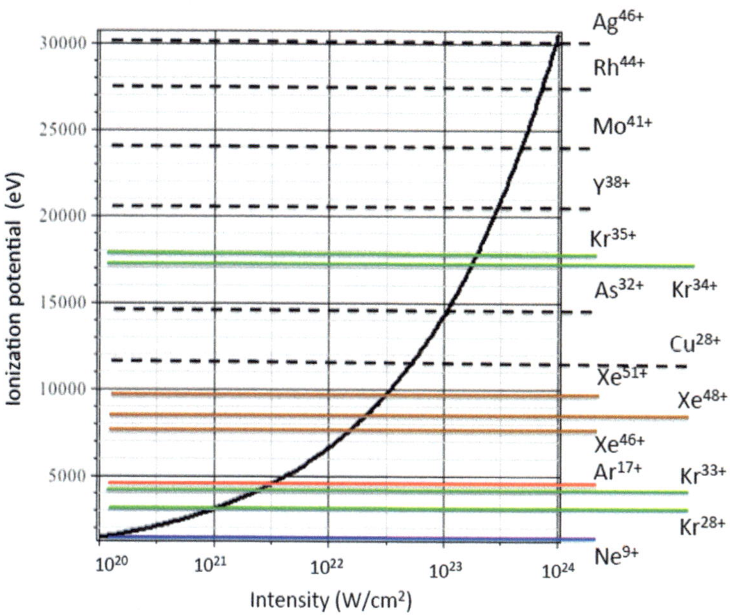

Fig. 8.1 (Color online) Ionization offset, (8.14), shown by a thick black line as a function of laser intensity. Ionization potentials of several highly charged ions are shown by horizontal lines, including neon (blue), argon (red), krypton (green) and xenon (brown). Ionization potentials of Xe^{52+} and Xe^{53+} exceed 40 keV and lay above the selected energy range (limited to ~30 keV). The figure allows estimating the charge numbers whose distribution should be calculated numerically in order to calibrate the intensity within some certain interval. If only noble gases are used, the most common species used in strong field experiments, two gaps in the laser energy determination appear, i.e., there exists a laser intensity range that can not be covered using only these atomic targets. The respective intervals of intensity are $\mathcal{I} \approx 3 \times 10^{22} - 2 \times 10^{23}$ W/cm^2 and $\mathcal{I} > 2.5 \times 10^{23}$ W/cm^2. These gaps can be filled using other elements than noble gases, e.g. metals. Dashed lines show ionization potentials of several metals with only one electron left in the ground $1s$ state. Note that in this plot the notation A^{N+} refers to the ionization potential of the respective ion, so that after one additional electron is removed, an ion with a charge $z = N + 1$ is generated

As is typical for logarithmic asymptotics, the denominator in the right-hand side (RHS) of (8.13) is a numerically large value of the order of 10. This gives the offset condition in the form (cf. [39]) $F^* \simeq 0.05$ or:

$$I_p^* \simeq \frac{1}{2}(20E_0)^{2/3} \approx 52.2\mathcal{I}^{1/3}. \tag{8.14}$$

In Fig. 8.1 we show the dependence, (8.14), versus laser intensity. Several ionization potentials of noble gases and metals are shown by horizontal solid and dashed lines, respectively.

Note that the logarithmic factor in the denominator of (8.13) grows with intensity via the growth of the ionization potential and therefore the value of K_0. This leads to a further suppression of the threshold value F^* at higher laser intensities. Solving (8.13) numerically one may find that $F^* \approx 0.08$ for the ground state of hydrogen, while for Ar^{15+}, with $I_p = 918.3$ eV, $F^* \approx 0.03$ ($\sim 9.7 \times 10^{18}$ W/cm^2) [26, 27]. This leads to an important conclusion simplifying our calculations: ionic states with higher ionization potentials are efficiently ionized at lower reduced fields, which makes the sub-barrier tunneling mechanism of ionization more appropriate than in the case of neutral atoms where one may enter the barrier suppression regime having $F^* \simeq 0.1$. For model analytic formulas describing multiple ionization in this intermediate regime see a recent paper [40].

The relativistic generalization of this result, which becomes quantitatively important at intensities exceeding 10^{24} W/cm^2, can be obtained by replacing (8.12) by its relativistic counterpart. The structure of the tunneling exponent remains the same with the difference that the characteristic field, (8.3c), is now determined as [30]

$$F_{\rm ch} = \frac{[\sqrt{3}\xi(I_p)]^3}{1+\xi^2(I_p)} c^3, \tag{8.15}$$

with $\xi^2(I_p)$ given by (8.4a). A calculation similar to that given above for the nonrelativistic limit, leads to an implicit formula for the relativistic offset I_p^*

$$\mathcal{I} = 1.57 \times 10^8 \frac{\xi^6(I_p^*)}{1+\xi^2(I_p^*)}, \tag{8.16}$$

which coincides with (8.14) for $I_p^* \ll c^2$. Accounting for the second term in the expansion of (8.4a) in powers of I_p/c^2, we obtain instead of (8.14):

$$I_p^* \approx 52.2\mathcal{I}^{1/3}\left(1 + 4.63 \times 10^{-4}\mathcal{I}^{1/3}\right). \tag{8.17}$$

The relativistic correction in (8.17) remains small even at 10^{24} W/cm^2, so that the two curves calculated along (8.14) and (8.17) stay visually indistinguishable for the intensity interval showed in Fig. 8.1. Thus, the nonrelativistic approximation for the ionization rate continues quantitatively correct at these intensities. A considerable difference between relativistic and nonrelativistic rates becomes apparent only at $\mathcal{I} \simeq 10^{26}$ W/cm^2. This remarkable extension of the nonrelativistic picture for the particular case of tunneling in a low frequency electromagnetic field deserves to be highlighted (see also a discussion in [30], where a comparison between the two above discussed rates is treated in a different way).

8.2.4 Effect of the Intensity Space Distribution

The estimate presented in (8.14) allows for deriving a universal approximate expression for the charge distribution of ions produced in the laser focus. In order to further proceed, we assume that an atom located at a point **r** will be ionized up to the charge corresponding to the value of ionization potential given by (8.14), which in turn depends on the local maximal value of intensity $\mathcal{I}(\mathbf{r})$. This model presumes that the pulse duration is sufficiently long as to strip out all levels with lower I_ps before the field reaches its maximum value. In the next section we verify the validity of this approximation by solving numerically a system of rate equations (8.11a)–(8.11c) for different ionic species. For the time being and for simplicity, we consider I_p as a continuous value. Then, the number of ions with ionization potentials in the interval $\mathrm{d}I_p$ is given by

$$\mathrm{d}N = -n_0 \frac{\mathrm{d}V}{\mathrm{d}I_p} \mathrm{d}I_p = 3n_0 \frac{I_p^2}{I_{pm}^{*3}} \frac{\mathrm{d}V}{\mathrm{d}\delta} dI_p, \quad \delta = \frac{\mathcal{I}}{\mathcal{I}_m}. \tag{8.18}$$

Here $\mathcal{I}_m$ is the peak intensity value in the focus and I_{pm}^* is the corresponding offset ionization potential given by (8.14). The space volume with intensity equal or exceeding $\mathcal{I}$ is denoted as $V(\delta)$:

$$V(\delta) = \int\limits_{\delta \le \delta'(\mathbf{r}) \le 1} \mathrm{d}^3 r. \tag{8.19}$$

For the simplest case of a fundamental Gaussian beam symmetric with respect to the x axis, we can write:

$$\mathcal{I}(r_\perp, z) = \frac{\mathcal{I}_m}{1 + x^2/z_R^2} \exp\left\{ -\frac{2r_\perp^2}{w_0^2(1 + x^2/z_R^2)} \right\}, \tag{8.20}$$

with $z_R = \pi w_0^2/\lambda$ being the Rayleigh length and w_0 the focal waist. Thus, a trivial calculation gives

$$V(\delta) = \frac{4\pi^2}{3} \frac{w_0^4}{\lambda} \left\{ \frac{1}{6} y^3 + y - \operatorname{atan}(y) \right\}, \quad y = \sqrt{\frac{1}{\delta} - 1} = \sqrt{\frac{\mathcal{I}_m}{\mathcal{I}} - 1}, \tag{8.21}$$

so that the distribution in ionization potentials (8.18) takes the form:

$$f(I_p, \mathcal{I}_m) \equiv \frac{\mathrm{d}N}{\mathrm{d}I_p} = \frac{\pi^2 n_0 w_0^4}{\lambda I_p} \sqrt{\frac{\mathcal{I}_m}{\mathcal{I}} - 1} \left(2 + \frac{\mathcal{I}_m}{\mathcal{I}} \right). \tag{8.22}$$

Note that the distribution is divergent at $I_p \to 0$ due to the formally unlimited focal volume. This makes, however, no difficulty for practical calculations, as the effective

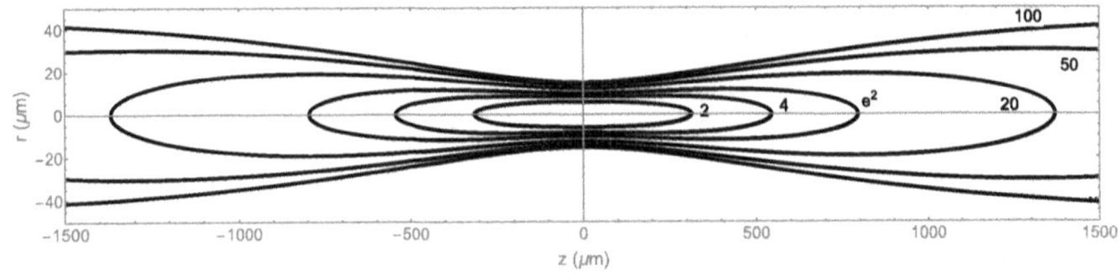

Fig. 8.2 Iso-intensity contours at the focus of a Gaussian-shaped laser beam. The labels refer to the ratio of the laser intensities at the center of the focus to the one at a given contour. The focal waist is $w_0 = 10\ \mu\text{m}$

volume is restricted either by the minimal value of the ionization potential of a neutral gas or by the geometry of the TOF detector [41].

In Fig. 8.2 we plot the iso-intensity contours at the focus of a Gaussian-shaped laser beam. We used a focal waist is $w_0 = 10\ \mu\text{m}$ ($z_R = 314\ \mu\text{m}$ for a $\lambda = 1\ \mu\text{m}$) and the labels refer to the ratio of the laser intensities at the center of the focus and at a given contour. For instance, if the value of intensity at focus is 10^{21} W/cm^2, the intensity at a distance of around 1.2 mm in the z-direction would be $\sim 5 \times 10^{19}$ W/cm^2.

Figure 8.3 shows the distribution (8.22) for several values of the peak focus intensity, as a function of the effective ion charge $z_{\text{eff}} = \sqrt{2I_p}$, equal to the charge of a hydrogen-like ion with identical ionization potential. Values of z_{eff} for several charge states in argon, krypton, and xenon are indicated by vertical lines to reproduce realistic discrete charge distributions. In order to approximate a relative amount of different charge states one has to look at intersections of vertical lines with the respective distribution curve. As an example, for ionization of krypton at $\mathcal{I} = 10^{24}$ W/cm^2 the number of Kr^{35+} and Kr^{36+} ions in the focus is expected to be approximately 3.5–4.0 orders of magnitude less than that of Kr^{29+}–Kr^{34+} ions. Instead, for ionization of xenon at $\mathcal{I} = 10^{23}$ W/cm^2 the number of Xe^{47+} ions is expected to be roughly three times more than that of Xe^{52+}, while Xe^{53+} and Xe^{54+} will not be produced at this intensity.

More precisely, the number of A^{N+} ions produced in the focus at a given $\mathcal{I}_m$ is obtained by calculating the area under the respective curve between the ionization potentials of $A^{(N-1)+}$ and A^{N+}. Using (8.21) this number can be explicitly expressed as:

$$N(A^{N+}) = n_0 \left\{ V(I_p(A^{(N-1)+})) - V(I_p(A^{N+})) \right\} \approx$$

$$\approx \frac{\pi^2 n_0 w_0^4}{\lambda} \sqrt{\frac{\mathcal{I}_m}{\mathcal{I}} - 1} \left(2 + \frac{\mathcal{I}_m}{\mathcal{I}}\right) \frac{\Delta I_p}{I_p}, \tag{8.23}$$

where $\Delta I_p = I_p(A^{(N-1)+}) - I_p(A^{N+}))$, $I_p = I_p(A^{(N-1)+})$ and $\mathcal{I} = \mathcal{I}(I_p)$ is calculated from (8.14). For the case of a fully stripped atom, i.e. for $n = Z$, the value of $I_p(A^{N+})$ does not exist, therefore we should modify (8.23) by replacing $I_p(A^{N+})$

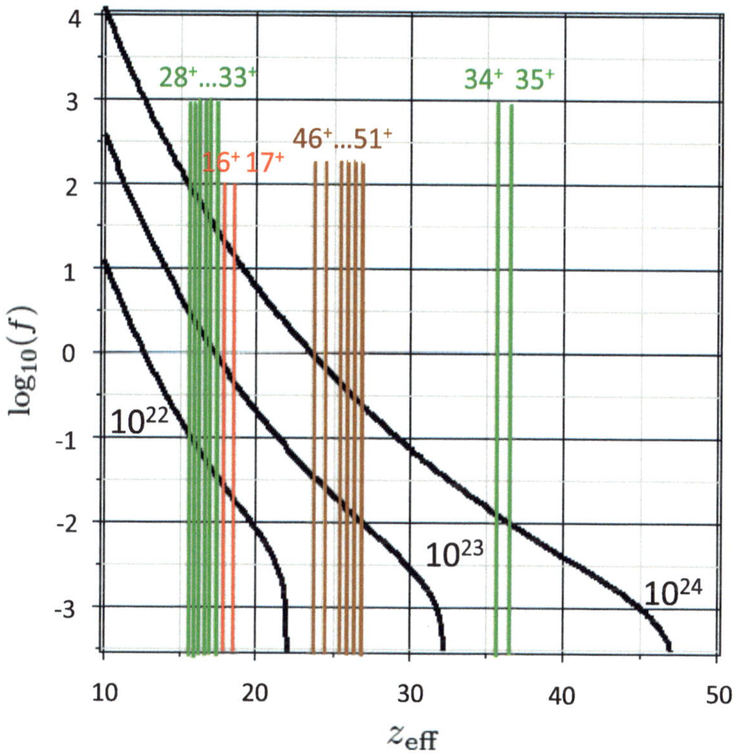

Fig. 8.3 (Color online) Charge distributions, (8.22) (in logarithmic scale,) calculated for $\mathcal{I}_m = 10^{22}$ W/cm^2, 10^{23} W/cm^2 and 10^{24} W/cm^2 as functions of the effective charge $z_{\text{eff}} = \sqrt{2I_p}$. The distributions are shown by thick black lines with the values of intensity indicated near each curve. Selected charge states of argon, krypton and xenon are shown by vertical red, green and brown lines respectively. The value at the intersection of vertical lines with the distributions approximately indicates the relative amount of the respective charge state at a given intensity

by I^*_{pm}. In particular, this formula allows for estimating the absolute number of ions $N(A^{N+})$ produced in the focus, which is instrumental to check the experimental feasibility to detect these ions. As an example, taking krypton at an intensity exceeding by 10% the threshold intensity for Kr^{35+} ($\sim 3 \times 10^{23}$ W/cm^2), using $I_p(\text{Kr}^{35+}) \approx 17936$ eV [44] and, assuming $n_0 = 10^{14}$ cm^{-3}, $\lambda = 1$ μm, and $w_0 \simeq 2\lambda$ we estimate $N(\text{Kr}^{36+}) \approx 80$, which should be amply sufficient for detection.

8.3 Numerical Calculations

In this section, we numerically validate the theory introduced in the previous Sections. In order to find the distribution of ionic charge states during and after the interaction with intense laser radiation, we solve a set of equations (8.11a)–(8.11c) using the nonrelativistic static rate, (8.9a), for argon, krypton and xenon in different ranges of laser intensities. We employ an adaptive stepsize Runge-Kutta scheme, based on the Numerical Recipes (for more details see [43]).

In a fully accurate calculation, one should start the simulation from a neutral atom and take into account all possible trees of ionic states which develop starting from this initial condition. A simple analysis shows, however, that the number of possible pathways grows enormously as the maximal charge number increases. Even assuming that (a) there is no ionization from inner shells until the outer shell is not fully striped out of electrons, (b) ionization proceeds without excitation of the residual remaining ion and (c) one neglects the fine structure of electronic terms, we obtain that the number of pathways for ionization of argon is equal to $P = 1$ for initial states Ar^{N+} with $N \geq 14$, $P = 6$ for $N = 12$, $P = 28$ for $N = 8$, etc. Calculation of P for the initial configuration $1s^2 2s^2 2p^6$ (corresponding to neutral neon, Ar^{8+}, etc.), based on these simplifications, is presented in Appendix A.

The analytical estimates of previous Sections allow to dramatically reduce the complexity of the problem. This is so by assuming that ionization starts from ionic charge states whose ionization potential can be estimated from (8.14), while all the outermost levels have been quickly ionized before the peak value of intensity has been attained. Thus, our numerical calculation intends to show the ionization dynamics of levels with ionization potentials not very much dissimilar of those given by (8.14). Such computation will verify the accuracy of the analytic estimate and show how quickly ionization saturation can be achieved at a fixed laser intensity.

For clarity, we suppose that an experiment aims probing laser intensities in a given fix interval, e.g. $\mathcal{I} = 10^{21} - 10^{23}$W/cm^2 at the center of the focus (see Fig. 8.2). According to Fig. 8.1, in this interval one would expect significant production of Ar^{18+} (fully stripped argon) and Kr^{28+}–Kr^{34+} ions. Ionic states with lower ionization potentials will be quickly ionized during the pulse intensity growth, while the probability of production of Kr^{35+}–Kr^{36+} will remain negligibly small in this range of intensities. In order to check these qualitative predictions, we solve the associated systems of rate equations for Ar and for a spatially homogeneous laser pulse linearly polarized ($\rho = 0$). In this case $F(t)$ takes the form:

$$F(t) = E_L f(t) \cos(\omega t), \tag{8.24}$$

where the pulse envelope is defined as $f(t) = \sin^2\left(\frac{\omega t}{2n_p}\right)$, with n_p the total number of optical cycles. In our simulations we use $\omega = 0.0455$ a.u. and $n_p = 10$, that correspond to a laser wavelength $\lambda = 1$ μm and a total pulse length $T \sim 33$ fs, respectively.

We consider the following assumptions:

(a) The initial state is chosen to have considerably lower ionization potential than $I_p^*(\mathcal{I}_{\max})$, but high enough to minimize the number of rate equations in the system. For actual calculations, we set up for the initial condition a gas of A^{k+} with k determined from (8.14) with intensity 2 orders in magnitude below the peak intensity under consideration. As an example, considering ionization by pulses with

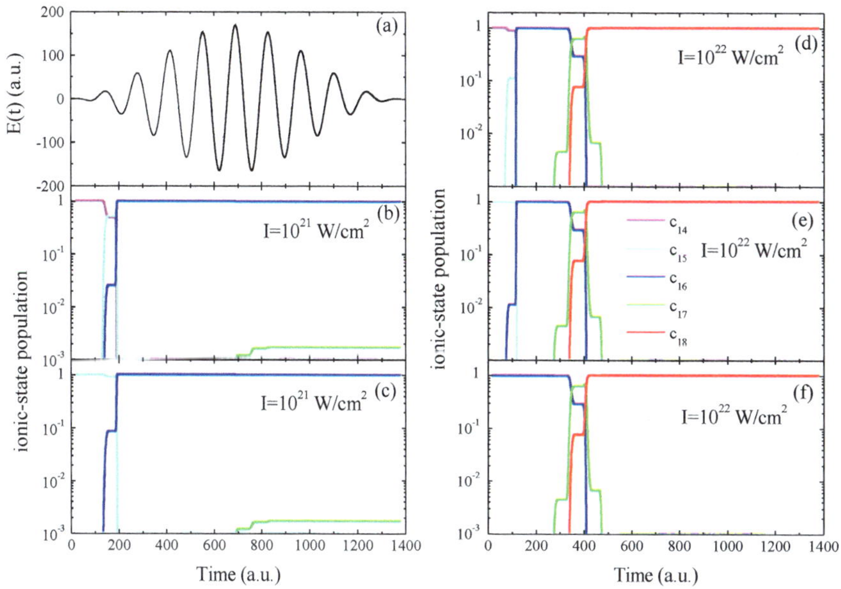

Fig. 8.4 **a** Time profile of the laser electric field for an equivalent laser intensity $\mathcal{I}_m = 10^{21}$ W/cm^2; **b–f** populations c_n for Ar^{N+} ions as a function of time calculated by solving numerically the system of rate equations (B25)–(B29). For (**b**) and (**c**) the laser intensity is set to $\mathcal{I}_m = 10^{21}$ W/cm^2 and for (**d**)–(**f**) to $\mathcal{I}_m = 10^{22}$ W/cm^2. The initial conditions are **b** $c_{14}(0) = 1,\ c_{15}(0) = \cdots = c_{18}(0) = 0$; **c** $c_{14}(0) = 0,\ c_{15}(0) = 1,\ c_{16}(0) = \cdots = c_{18}(0) = 0$; **d** $c_{14}(0) = 1,\ c_{15}(0) = \cdots = c_{18}(0) = 0$; **e** $c_{14}(0) = 0,\ c_{15}(0) = 1,\ c_{16}(0) = \cdots = c_{18}(0) = 0$ and **f** $c_{14}(0) = c_{15}(0) = 0,\ c_{16}(0) = 1,\ c_{17}(0) = c_{18}(0) = 0$

$\mathcal{I}_{max} = 10^{21}$ W/cm^2 we assume a first set of calculations setting $c_{14}(0) = 1$ for argon (see below for more details). In order to justify the validity of such model initial conditions, we ran several calculations with different initial ionic states and confirmed that the final ionic states distribution remains insensitive to the initial choice.

(b) Ionization is not followed by excitation of the residual ion. This assumption is largely justified by the deep tunnel regime of ionization we consider, when the Keldysh parameter in (8.6) remains well below 0.1. However, this assumption does not imply that an electron with a slightly higher ionization potential can be removed before that with a smaller one, provided the difference between the ionization potentials is relatively not too high. This may happen for shells containing s and p-electrons, as shown in Appendix A.

Assumptions (a, b) further reduce the number of rate equations in the system. For simulations, we take the data on ionization potentials and (l, m) configurations of ionic levels in argon, krypton and xenon from the fundamental works by Saloman [42, 44, 45]. We first compute the dependence on time of the populations $c_{14} \ldots c_{18}$ for argon in the field (8.24) with two different peak intensities: 10^{21} and 10^{22} W/cm^2. The results are shown in Fig. 8.4, where the populations $c_{14} \ldots c_{18}$ are present as functions of time for different initial conditions assuming that the pre-ionized gas consists

Fig. 8.5 Populations c_n for Ar^{N+} ions at the end of the laser pulse as functions of the laser peak intensity. The initial conditions are **a** $c_{14}(0) = 1,\ c_{15}(0) = \cdots = c_{18}(0) = 0$, **b** $c_{14}(0) = 0,\ c_{15}(0) = 1,\ c_{16}(0) = \cdots = c_{18}(0) = 0$ and **c** $c_{14}(0) = c_{15}(0) = 0,\ c_{16}(0) = 1,\ c_{17}(0) = c_{18}(0) = 0$. Arrows mark the saturation intensities which have to be compared with those extracted from (8.14) or Fig. 8.1. Note that here we point out the saturation intensity needed to generate fully stripped ions, i.e. without any electron present, which correspond, in notations of Fig. 8.1, to the ionization potential of A^{n-1+} ions

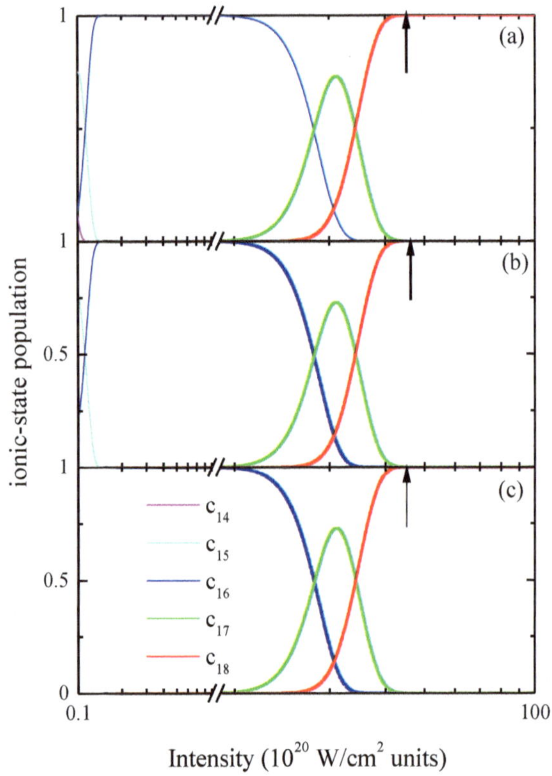

of Ar^{14+} (Fig. 8.4b–d), Ar^{15+} (Fig. 8.4c–e) and Ar^{16+} (Fig. 8.4f), respectively. In Fig. 8.4b the laser intensity is set to 10^{21} W/cm^2, meanwhile in Fig. 8.4d–f we set it to 10^{22} W/cm^2.

In Fig. 8.5 we plot the ionic states of Ar at the end of the laser pulse as a function of the laser intensity in a range of 10^{19}–10^{22} W/cm^2. We make use of 3 different initial conditions, namely (i) $c_{14}(0) = 1,\ c_{15}(0) = \cdots = c_{18}(0) = 0$ (Fig. 8.4a); (ii) $c_{14}(0) = c_{16}(0) = \cdots = c_{18}(0) = 0,\ c_{15}(0) = 1$ (Fig. 8.4b) and (iii) $c_{14}(0) = c_{15}(0) = 0,\ c_{16}(0) = 1,\ c_{17}(0) = c_{18}(0) = 0$ (Fig. 8.4c) (see Appendix B for more details). Here the ionization rates $w(\nu_z, l, m; t)$ are obtained from (8.9a) with the effective principal quantum numbers $\nu_{14} = 1.89$, $\nu_{15} = 1.95$, $\nu_{16} = 0.98$ and $\nu_{17} = 1.00$ calculated from (8.8), using the ionization potentials taken from [42].

A couple of conclusions can be drawn at this point. First, the results shown in Figs. 8.4 and 8.5 fully support our assumption (a): the final distribution of charge states appears indeed insensitive to the initial state, provided its ionization potential is low enough. In this case solutions corresponding to different initial states appear to be almost indistinguishable. Second, and even more relevant and decisive, the different curves of Figs. 8.4 and 8.5 show the reliability of the analytic estimations given by (8.14). The charge state Ar^{16+} is reached at $\mathcal{I} \approx 1.3 \times 10^{19}$ W/cm^2, and Ar^{18+}—

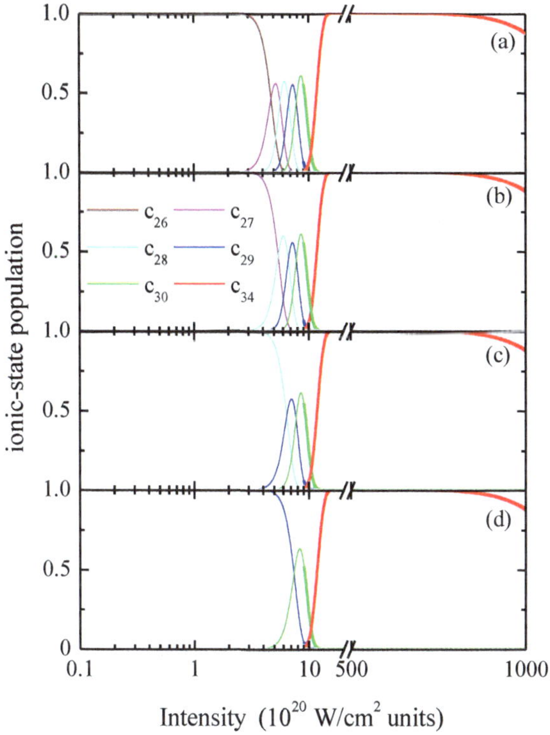

Fig. 8.6 Populations c_n for Kr^{N+} ions at the end of the laser pulse as functions of the laser peak intensity for the case (**a**) (see Appendix B for details). The initial conditions are **a** $c_{26}(0) = 1,\ c_{27}(0) = \cdots = c_{33}(0) = 0$, **b** $c_{26}(0) = 0,\ c_{27}(0) = 1,\ c_{28}(0) = \cdots = c_{34}(0) = 0$, **c** $c_{26}(0) = c_{27}(0) = 0,\ c_{28}(0) = 1,\ c_{29}(0) = \cdots = c_{34}(0) = 0$ and **d** $c_{26}(0) = \cdots = c_{28}(0) = 0,\ c_{29}(0) = 1,\ c_{30}(0) = \cdots = c_{34}(0) = 0$. For clarity, only the relevant ionic states populations are plotted. The saturation intensity for the Kr^{34+} ions lies beyond $\mathcal{I} = 10^{23}$ W/cm^2 [see Fig. 8.1 and (8.14)]

at $\mathcal{I} \approx 4 \times 10^{21}$ W/cm^2, in a reasonable agreement with the analytic estimates. The latter assumes that, as the value of intensity grows, populations of charge states evolve from 0 to 1 and then back to 0 in small intensity intervals, so that the coefficients c_n can be approximated by step functions. As is seen from the plots of Fig. 8.4, populations of the levels behave almost as Heaviside step functions, justifying the validity of the latter approximation.

A similar calculation was performed for krypton taking as initial condition ionic states between Kr^{26+} and Kr^{31+} and in an intensity range between 10^{19} W/cm^2 and 10^{23} W/cm^2 (see caption of Fig. 8.6 for more details). In the first case we have taken into account only one pathway corresponding to the removal of the outermost electron always, i.e. the p-electrons are removed first. This pathway is shown by red arrows on Fig. 8.9. In the second case all six relevant pathways are taken into account. The systems of rate equations for this case are given in Appendix B. The corresponding distributions of the relevant ionic populations are shown on Figs. 8.6 and 8.7. We can observe that, as in the case of Ar, the final ionic state is independent of the initial condition. Interestingly, this behavior remains even when (i) we include excited states (case b) and (ii) we employ completely different pathways to reach the same final state (compare the set of rate equations (B30)–(B38) with (B39)–(B47), Appendix B).

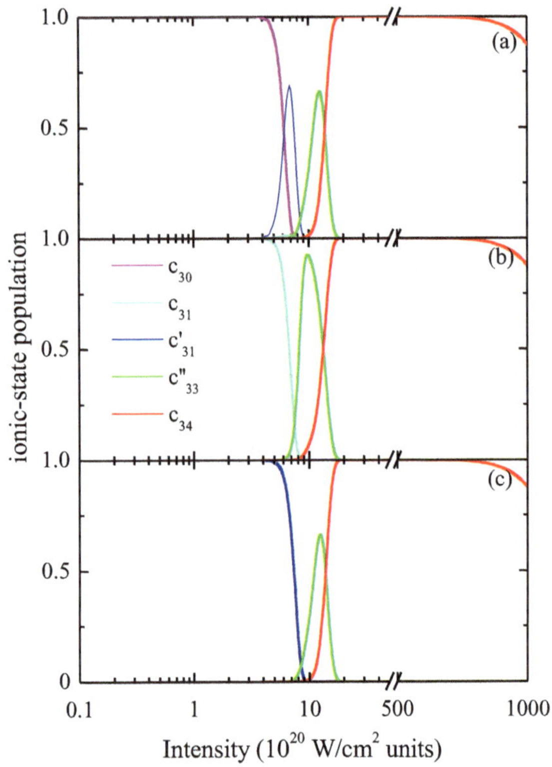

Fig. 8.7 Populations c_n for Kr^{N+} ions at the end of the laser pulse as functions of the laser peak intensity for the case (**b**) (see Appendix B for details). The initial conditions are **a** $c_{30}(0) = 1,\ c_{31}(0) = \cdots = c_{34}(0) = 0$, **b** $c_{30}(0) = 0,\ c_{31}(0) = 1,\ c'_{31}(0) = \cdots = c_{34}(0) = 0$ and **c** $c_{30}(0) = c_{31}(0) = 0,\ c'_{31}(0) = 1,\ c_{32}(0) = \cdots = c_{34}(0) = 0$. For clarity, only the relevant ionic states populations are plotted. The saturation intensity for the Kr^{34+} ions lies beyond $\mathcal{I} = 10^{23}$ W/cm^2 [see Fig. 8.4 and (8.14)]

Finally, and for completeness, Fig. 8.8 shows the populations of Xe ions at the end of the laser pulse as a function of the laser intensity in a range 10^{21}–5×10^{24} W/cm^2. In Fig. 8.8a we start our simulations with $c_{50}(0) = 1,\ c_{51}(0) = \cdots = c_{54}(0) = 0$, in Fig. 8.8b with $c_{50}(0) = 0,\ c_{51}(0) = 1,\ c_{53}(0) = \cdots = c_{54}(0) = 0$ and in Fig. 8.8c $c_{50}(0) = c_{51}(0) = 0,\ c_{52}(0) = 1,\ c_{53}(0) = c_{54}(0) = 0$. As in the previous two cases, we observe a very good agreement between the saturation intensities obtained numerically with the analytical estimates derived from (8.14).

8.4 Conclusions and Outlook

In summary, we have demonstrated that using the strong dependence of the ionization off-set in complex multielectronic atoms on laser intensity the latter can be reliably estimated. The simple analytic estimates of Sect. 8.2 qualitatively agree with the full numerical results presented in Sect. 8.3. The former help identifying intervals of ionization potentials necessary to probe a certain range of laser intensities, but do not provide quantitative accuracy. Therefore, for a precise determination of the maximal

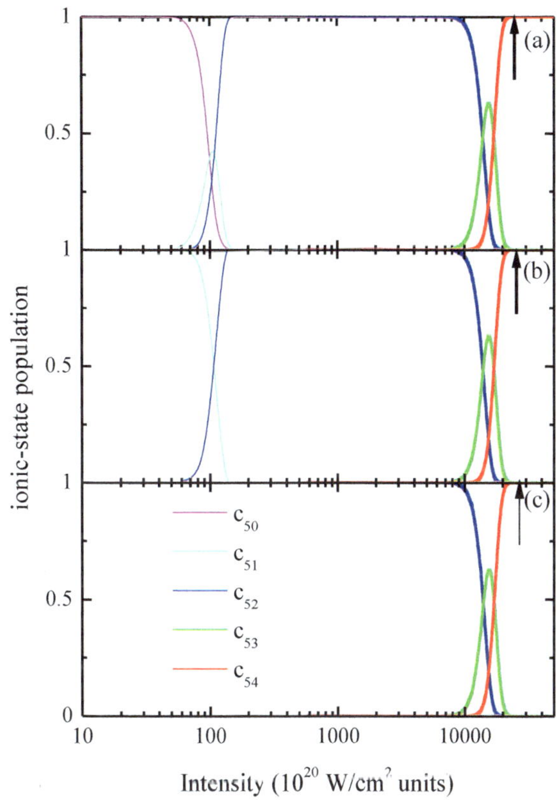

Fig. 8.8 Populations c_n for Xe^{N+} ions at the end of the laser pulse as functions of the laser peak intensity. The initial conditions are **a** $c_{50}(0) = 1,\ c_{51}(0) = \cdots = c_{54}(0) = 0$, **b** $c_{50}(0) = 0,\ c_{51}(0) = 1,\ c_{52}(0) = \cdots = c_{54}(0) = 0$ and **c** $c_{50}(0) = c_{51}(0) = 0,\ c_{52}(0) = 1,\ c_{53}(0) = c_{54}(0) = 0$. Arrows mark the saturation intensities which have to be compared with those extracted from (8.14). Note that here we point out the saturation intensity needed to generate bare ions, which correspond, in notations of Fig. 8.1, to the ionization potential of A^{n-1+} ions

laser intensity in the focus, numerical solution of the rate equations is required along the lines presented in Sect. 8.3 and Appendix B.

Two instrumental experimental-related issues will probably be faced on the way to an actual implementation of this diagnostics tool. First, a spatial distribution of laser intensity in the focus will lead to the volume effect analytically examined in Sect. 2.4 resulting in production of a very significant number of relatively low charged ions which can saturate the signal recorded by the TOF detector. This issue can be resolved by mechanically restricting the effective volume accessible for the detector or by deflecting ions with sufficiently low charge states in a magnetic field before these ions reach the detector (see e.g. [33]). Second, as discussed in Sect. 8.2 and shown by Fig. 8.1, there are considerable gaps in ionization potentials (and, correspondingly, an impossibility to estimate the laser intensity in these regions) which cannot be filled in by noble gases, which are the most common elements being employed as gas targets in strong field physics. Potential candidates, most likely highly charged ions of metals, such as Ag, Mo, Cu, etc., have to be examined in virtue of their feasibility for such an experiment in this laser intensity regions.

Acknowledgements We are deeply indebted to S. V. Popruzhenko for useful comment and suggestions. Authors acknowledge fruitful discussions with D. Batheja. Supported by the project High Field Initiative (CZ.02.1.01/0.0/0.0/15_003/0000449) from European Regional Development Fund (HIFI) and by the project Advanced research using high intensity laser produced photons and particles (CZ.02.1.01/0.0/0.0/16_019/0000789) from European Regional Development Fund (ADONIS). The results of the project LQ1606 were obtained with the financial support of the Ministry of Education, Youth and Sports as part of targeted support from the National Programme of Sustainability II. M. F. C acknowledges Prof. K. Yamanouchi for the invitation to participate in this project and to M. Abe for her exceptional assistance. The text of the following chapter is partially reproduced from [34], with kind permission from the American Physical Society.

Appendix A. Ionization Cascades

Starting from a given ground state of an atom or an ion, ionization can proceed along different pathways. Clearly, for multi electron atoms the total number of such paths grows very quickly with the atomic number. Most of them do not give any considerable contribution into the production of ionic states owing to the structure of the tunneling rate. The tunneling exponent in (8.9a) is maximal for the outermost electron which has the minimal ionization potential. This makes ionization of inner shells highly improbable before the outer shells have been stripped out, so that such ionization pathways can be safely discarded. However, when electrons are being removed from the same shell, ionization of lower s-levels may proceed with comparable or even higher probability than that of p-levels with lower ionization potentials. The reasons for that is (a) a smaller value of the asymptotic coefficients $C_{\nu l}^2$ in (8.9b) and (b) the factor $F^{|m|}$ in (8.9a), which is small for nonzero magnetic quantum numbers, owing to the condition $F \ll 1$. As an example, for Kr^{26+} with the ground state configuration $1s^2 2s^2 2p^6$, ionization of a p-electron with $I_p \approx 2929$ eV proceeds with probability comparable to that for an s-electron whose ionization potential is by $\Delta I_p \approx 235$ eV higher [44]. Indeed, taking $F \simeq F^* = 0.05$, one obtains that the tunneling exponent for the p-electron is $\exp(\Delta I_p / I_p F) \approx 5$ times higher than that for the s-electron. At the same time, for the latter $C_{\nu 0}^2 = 1.088,\ B_{00} = 1$, while for the p-electron $C_{\nu 1}^2 = 0.315$ and $B_{10} = 3$, $B_{1\pm 1} = 3/2$, and finally the factor $F^{|m|}$ gives 0.05 for $m = \pm 1$. Thus, the p-state rate averaged over the magnetic quantum number appears only 5 times greater than that of the s-state. For partially stripped p-shells the difference in ionization potentials appears to be even smaller, so that for configurations $2s^2 2p$ or $2s^2 2p^2$, the s- and p-rates are almost equal. These estimates show that the described sub-manifold of pathways may play an essential role in the ionization dynamics.

An example of the structure of levels is shown on Fig. 8.9 for the initial configuration $1s^2 2s^2 2p^6$, which corresponds to the neutral neon, Ar^{10+}, Kr^{26+}, etc.

For the initial configuration shown on Fig. 8.9 the total number of pathways is equal to 28. For the $1s^2 2s^2 2p^2$ configuration only 5 relevant pathways left; starting from the $1s^2 2s^2$ configuration ionization proceeds along the unique pathway.

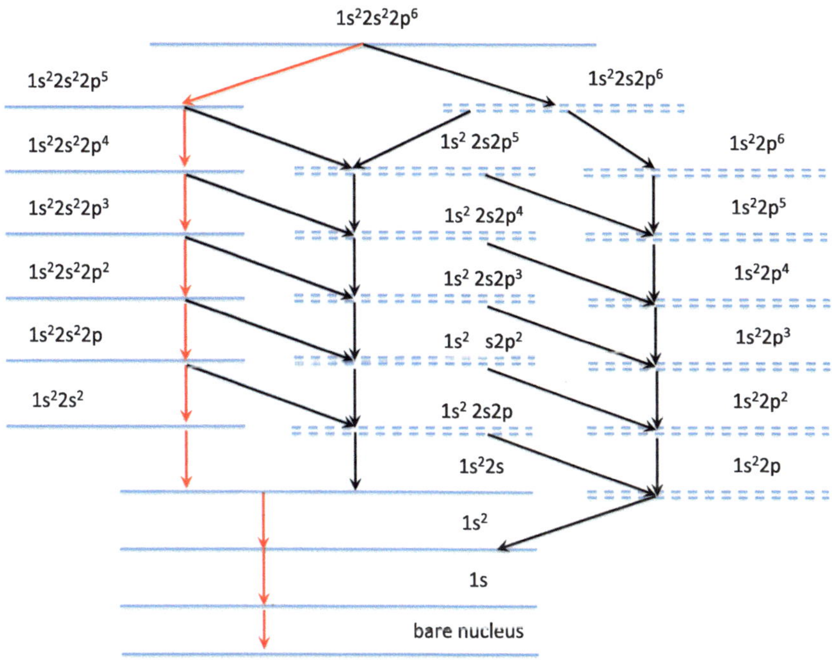

Fig. 8.9 Substructure of levels and ionization pathways making the main contribution into sequential multiple ionization of atomic systems prepared in the ground $1s^2 2s^2 2p^6$ state. Ground and excited states are shown by blue solid and dashed lines correspondingly. Most of the excited states consist of several sub-levels with different values of the full angular momentum J. The ionization pathway which involves only the ground states is shown by red arrows, all other pathways—by black arrows

Appendix B. Systems of Rate Equations

We solve numerically the system of rate equations for argon in the interval of intensities $\mathcal{I}_m = 10^{19}$–10^{22} W/cm^2 and for krypton between $\mathcal{I}_m = 10^{19}$–10^{23} W/cm^2 using an adaptive stepsize Runge-Kutta scheme [43]. We start with the simplest configuration $1s^2 2s^2$ for argon (Ar^{14+}, $I_p \approx 855$ eV). The value of I_p is well below that of 8.14 for 10^{20}W/cm^2 which is $I_p^* \approx 1420$ eV. For this initial configuration, only one relevant pathway contributes (see Fig. 8.9 and Appendix A). The resulting system of rate equations is therefore particularly simple and reads:

$$\frac{dc_{14}}{dt} = -2c_{14}w(\nu_{14}, 0, 0; t), \tag{B25}$$

$$\frac{dc_{15}}{dt} = 2c_{14}w(\nu_{14}, 0, 0; t) - c_{15}w(\nu_{15}, 0, 0; t), \tag{B26}$$

$$\frac{dc_{16}}{dt} = c_{15}w(\nu_{15}, 0, 0; t) - 2c_{16}w(\nu_{16}, 0, 0; t), \tag{B27}$$

$$\frac{dc_{17}}{dt} = 2c_{16}w(\nu_{16}, 0, 0; t) - c_{17}w(\nu_{17}, 0, 0; t), \tag{B28}$$

$$\frac{dc_{18}}{dt} = c_{17}w(\nu_{17}, 0, 0; t). \tag{B29}$$

Coefficients 2 in (B25)–(B28) are due to the presence of two equivalent electrons in the sub-shell.

For the interval of intensities used for Kr, the system of rate equations has to include p-states of the $2p$ shell. In order to simplify the calculations, we consider two cases: (a) the $1s^2 2s^2 2p^6$ state as initial configuration, with only the most probably pathway accounted for (shown on Fig. 8.9 by red arrows) and (b) the $1s^2 2s^2 2p^2$ as initial configuration with all relevant pathways accounted for (shown on Fig. 8.9 by red and black arrows). For case (a) the system explicitly reads:

$$\frac{dc_{26}}{dt} = -2c_{26}\{w(\nu_{26}, 1, 0; t) + 2w(\nu_{26}, 1, \pm 1; t)\}, \tag{B30}$$

$$\begin{aligned} \frac{dc_{27}}{dt} &= 2c_{26}\{w(\nu_{26}, 1, 0; t) + 2w(\nu_{26}, 1, \pm 1; t)\} \\ &- \frac{5}{3}c_{27}\{w(\nu_{27}, 1, 0; t) + 2w(\nu_{27}, 1, \pm 1; t)\}, \end{aligned} \tag{B31}$$

$$\begin{aligned} \frac{dc_{28}}{dt} &= \frac{5}{3}c_{27}\{w(\nu_{27}, 1, 0; t) + 2w(\nu_{27}, 1, \pm 1; t)\} \\ &- \frac{4}{3}c_{28}\{w(\nu_{28}, 1, 0; t) + 2w(\nu_{28}, 1, \pm 1; t)\}, \end{aligned} \tag{B32}$$

$$\begin{aligned} \frac{dc_{29}}{dt} &= \frac{4}{3}c_{28}\{w(\nu_{28}, 1, 0; t) + 2w(\nu_{28}, 1, \pm 1; t)\} \\ &- c_{29}\{w(\nu_{29}, 1, 0; t) + 2w(\nu_{29}, 1, \pm 1; t)\}, \end{aligned} \tag{B33}$$

$$\begin{aligned} \frac{dc_{30}}{dt} &= c_{29}\{w(\nu_{29}, 1, 0; t) + 2w(\nu_{29}, 1, \pm 1; t)\} \\ &- \frac{2}{3}c_{30}\{w(\nu_{30}, 1, 0; t) + 2w(\nu_{30}, 1, \pm 1; t)\}, \end{aligned} \tag{B34}$$

$$\begin{aligned} \frac{dc_{31}}{dt} &= \frac{2}{3}c_{30}\{w(\nu_{30}, 1, 0; t) + 2w(\nu_{30}, 1, \pm 1; t)\} \\ &- \frac{1}{3}c_{31}\{w(\nu_{31}, 1, 0; t) + 2w(\nu_{31}, 1, \pm 1; t)\}, \end{aligned} \tag{B35}$$

$$\frac{dc_{32}}{dt} = \frac{1}{3}c_{31}\{w(\nu_{31}, 1, 0; t) + 2w(\nu_{31}, 1, \pm 1; t)\} - 2c_{32}w(\nu_{32}, 0, 0; t), \tag{B36}$$

$$\frac{dc_{33}}{dt} = 2c_{32}w(\nu_{32}, 0, 0; t) - c_{33}w(\nu_{33}, 0, 0; t), \tag{B37}$$

$$\frac{dc_{34}}{dt} = c_{33}w(\nu_{33}, 0, 0; t) - 2c_{34}w(\nu_{34}, 0, 0; t). \tag{B38}$$

The system can be safely truncated by (B38), as the ionization potential of Kr^{34+}, $I_p = 17296$ eV, is too high to expect any considerable ionization below 10^{23} W/cm^2 (see Figs. 8.1 and 8.2). The effective principal quantum numbers ν_z are calculated using data from [44].

For case (b) we take into account all relevant pathways (see red and black arrows in Fig. 8.9) up to the same ionic state as the one used in (a). As a result, excited states of two types, $1s^2 2s 2p^n$ and $1s^2 2p^n$ with $n = 1, 2$ enter in the calculation. We denote the values corresponding to these two sets of excited states by one and two primes, respectively. Then the system of rate equations reads:

$$\frac{dc_{30}}{dt} = -c_{30}\{\frac{2}{3}[w(\nu_{30}, 1, 0; t) + 2w(\nu_{30}, 1, \pm 1; t)] + 2w(\nu'_{30}, 0, 0; t)\}, \tag{B39}$$

$$\frac{dc_{31}}{dt} = -c_{31}\{\frac{1}{3}[w(\nu_{31}, 1, 0; t) + 2w(\nu_{31}, 1, \pm 1; t)] + 2w(\nu'_{31}, 0, 0; t)\} + c_{30}\frac{2}{3}[w(\nu_{30}, 1, 0; t) + 2w(\nu_{30}, 1, \pm 1; t)], \tag{B40}$$

$$\frac{dc'_{31}}{dt} = -c'_{31}\{\frac{2}{3}[w(\nu'_{31}, 1, 0; t) + 2w(\nu'_{31}, 1, \pm 1; t)] + w(\nu''_{31}, 0, 0; t)\} + 2c_{30}w(\nu'_{30}, 0, 0; t), \tag{B41}$$

$$\frac{dc_{32}}{dt} = -2c_{32}w(\nu_{32}, 0, 0; t) + c_{31}\{\frac{1}{3}[w(\nu_{31}, 1, 0; t) + 2w(\nu_{31}, 1, \pm 1; t)]\}, \tag{B42}$$

$$\frac{dc'_{32}}{dt} = -c'_{32}\{\frac{1}{3}[w(\nu'_{32}, 1, 0; t) + 2w(\nu'_{32}, 1, \pm 1; t)] + w(\nu''_{32}, 0, 0; t)\} + 2c_{31}w(\nu'_{31}, 0, 0; t) + c'_{31}\{\frac{2}{3}[w(\nu'_{31}, 1, 0; t) + 2w(\nu'_{31}, 1, \pm 1; t)]\}, \tag{B43}$$

$$\frac{dc''_{32}}{dt} = -\frac{2}{3}c''_{32}[w(\nu''_{32}, 1, 0; t) + 2w(\nu''_{32}, 1, \pm 1; t)] + c'_{31}w(\nu''_{31}, 0, 0; t), \tag{B44}$$

$$\frac{\mathrm{d}c_{33}}{\mathrm{d}t} = -c_{33}w(\nu_{33}, 0, 0; t) + 2c_{32}w(\nu_{32}, 0, 0; t) + \frac{1}{3}c'_{32}[w(\nu'_{32}, 1, 0; t) + 2w(\nu'_{32}, 1, \pm 1; t)], \tag{B45}$$

$$\frac{\mathrm{d}c''_{33}}{\mathrm{d}t} = -\frac{1}{3}c''_{33}[w(\nu'_{33}, 1, 0; t) + 2w(\nu'_{33}, 1, \pm 1; t)] + c'_{32}w(\nu''_{32}, 0, 0; t) + \frac{2}{3}c''_{32}[w(\nu''_{32}, 1, 0; t) + 2w(\nu''_{32}, 1, \pm 1; t)], \tag{B46}$$

$$\frac{\mathrm{d}c_{34}}{\mathrm{d}t} = c_{33}w(\nu_{33}, 0, 0; t) + \frac{1}{3}c''_{33}[w(\nu'_{33}, 1, 0; t) + 2w(\nu'_{33}, 1, \pm 1; t)] - 2c_{34}w(\nu_{34}, 0, 0; t). \tag{B47}$$

Table 8.1 Parameters for the first ionization cascade of Kr

Ionic state	e-configuration	I_p (eV)	ν	C^2	l	m	B_{lm}
Kr^{26+}	$1s^2 2s^2 2p^6$	2 929	1.840	0.315	1	0	3
Kr^{26+}	$1s^2 2s^2 2p^6$	2 929	1.840	0.315	1	±1	3/2
Kr^{27+}	$1s^2 2s^2 2p^5$	3 072	1.863	0.319	1	0	3
Kr^{27+}	$1s^2 2s^2 2p^5$	3 072	1.863	0.319	1	±1	3/2
Kr^{28+}	$1s^2 2s^2 2p^4$	3 238	1.882	0.321	1	0	3
Kr^{28+}	$1s^2 2s^2 2p^4$	3 238	1.882	0.321	1	±1	3/2
Kr^{29+}	$1s^2 2s^2 2p^3$	3 381	1.903	0.322	1	0	3
Kr^{29+}	$1s^2 2s^2 2p^3$	3 381	1.903	0.322	1	±1	3/2
Kr^{30+}	$1s^2 2s^2 2p^2$	3 584	1.910	0.325	1	0	3
Kr^{30+}	$1s^2 2s^2 2p^2$	3 584	1.910	0.325	1	±1	3/2
Kr^{31+}	$1s^2 2s^2 2p$	3 752	1.927	0.327	1	0	3
Kr^{31+}	$1s^2 2s^2 2p$	3 752	1.927	0.327	1	±1	3/2
Kr^{32+}	$1s^2 2s^2$	3 971	1.931	1.030	0	0	1
Kr^{33+}	$1s^2 2s$	4 108	1.956	1.029	0	0	1
Kr^{34+}	$1s^2$	17 296	0.981	0.990	0	0	1
Kr^{35+}	$1s$	17 936	0.995	0.995	0	0	1

Appendix C. Parameters for Krypton and Xenon

Here the values of I_p, ν, $C_{\nu l}^2$ and B_{lm} are given for the two ionization cascades in krypton described by (B30)–(B38) and (B39)–(B47) and for the xenon case, that can be simulated using the same set of rate equations (B25)–(B29), changing the respective ionic state populations and their associated parameters.

The first case of Kr corresponds to ionization of the outermost orbitals (red arrows on Fig. 8.9) and starts from the $1s^2 2s^2 2p^6$ state of Kr^{26+}. The parameters of the first ionization cascade are shown in the Table 8.1. The second one starts from the $1s^2 2s^2 2p^2$ state of Kr^{30+} and accounts all pathways (shown by red and black arrows on Fig. 8.9).

For this case one should take into account that several transitions may occur from/to a given state. These transitions are depicted and numbered in Fig. 8.10. The respective parameters are presented in Table 8.2.

Finally, we run the same system of rate equations (B25)–(B29) for xenon starting from the ground state configuration $1s^2 2s^2$ of the Xe^{50+} ion. The parameters are shown the Table 8.3.

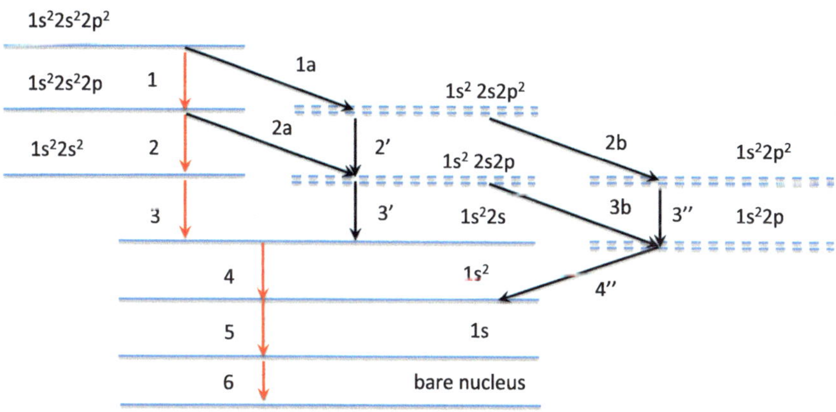

Fig. 8.10 Same as on Fig. 8.9 but for the initial ground state $1s^2 2s^2 2p^2$. See the text for details about the numeration of transitions and their parameters

Table 8.2 Parameters for the second ionization cascade of Kr (see text and Fig. 8.10 for more details)

Transition no	z	I_p (eV)	ν	C^2	l	m	B_{lm}
1	31	3 584	1.910	0.325	1	0	3
1	31	3 584	1.910	0.325	1	±1	3/2
1a	31	3 671	1.887	1.048	0	0	1
2	32	3 752	1.927	0.327	1	0	3
2	32	3 752	1.927	0.327	1	±1	3/2
2a	32	3 819	1.910	1.039	0	0	1
2'	32	3 732	1.932	0.327	1	0	3
2'	32	3 732	1.932	0.327	1	±1	3/2
2b	32	3 841	1.904	1.041	0	0	1
3	33	3 971	1.931	1.030	0	0	1
3'	33	3 904	1.949	0.329	1	0	3
3'	33	3 904	1.949	0.329	1	±1	3/2
3b	33	3 975	1.930	1.031	0	0	1
3"	33	3 865	1.958	0.330	1	0	3
3"	33	3 865	1.958	0.330	1	±1	3/2
4	34	4 109	1.956	1.020	0	0	1
4"	33	4 038	1.973	0.331	1	0	3
4"	33	4 038	1.973	0.331	1	±1	3/2
5	35	17 296	0.981	0.990	0	0	1
6	36	17 936	0.991	0.995	0	0	1

Table 8.3 Parameters for Xe

ionic state	e-configuration	I_p (eV)	ν	C^2	l	m	B_{lm}
Xe^{50+}	$1s^2 2s^2$	9 607	1.919	1.035	0	0	1
Xe^{51+}	$1s^2 2s$	9 812	1.936	1.028	0	0	1
Xe^{52+}	$1s^2$	40 272	0.974	0.986	0	0	1
Xe^{53+}	$1s$	41 300	0.962	0.979	0	0	1

References

1. C. Danson, D. Hillier, N. Hopps, D. Neely, High Power Laser Sci. Eng. **3**, e3 (2015). https://doi.org/10.1017/hpl.2014.52
2. M.D. Perry, D. Pennington, B.C. Stuart, G. Tietbohl, J.A. Britten, C. Brown, S. Herman, B. Golick, M. Kartz, J. Miller, H.T. Powell, M. Vergino, V. Yanovsky, Opt. Lett. **24**, 160 (1999). https://doi.org/10.1364/OL.24.000160
3. S.-W. Bahk, P. Rousseau, T.A. Planchon, V. Chvykov, G. Kalintchenko, A. Maksimchuk, G. Mourou, V. Yanovsky, Opt. Lett. **29**, 2837 (2004). https://doi.org/10.1364/OL.29.002837

4. V. Yanovsky, V. Chvykov, G. Kalinchenko, P. Rousseau, T. Planchon, T. Matsuoka, A. Maksimchuk, J. Nees, G. Cheriaux, G. Mourou, K. Krushelnick, Opt. Express **16**, 2109 (2008). https://doi.org/10.1364/OE.16.002109
5. J.-P. Chambaret, O. Chekhlov, G. Cheriaux et al., Proc. of SPIE **7721**, 77211D (2010). https://doi.org/10.1117/12.854687
6. A.V. Bashinov, A.A. Gonoskov, A.V. Kim, G. Mourou, A.M. Sergeev, Eur. Phys. J. Spec. Top. **223**, 1105 (2014). https://doi.org/10.1140/epjst/e2014-02161-7
7. E. Cartlidge, Science **359**, 382 (2018). https://doi.org/10.1126/science.359.6374.382
8. A.S. Pirozhkov, Y. Fukuda, M. Nishiuchi, H. Kiriyama, A. Sagisaka, K. Ogura, M. Mori, M. Kishimoto, H. Sakaki, N.P. Dover, K. Kondo, N. Nakanii, K. Huang, M. Kanasaki, K. Kondo, M. Kando, Opt. Express **25**, 20486 (2017). https://doi.org/10.1364/OE.25.020486
9. Z. Guo, L. Yu, J. Wang, C. Wang, Y. Liu, Z. Gan, W. Li, Y. Leng, X. Liang, R. Li, Opt. Express **26**, 26776 (2017). https://doi.org/10.1364/OE.26.026776
10. S. Weber et al., Matter Radiat. Extremes **2**, 149 (2017). https://doi.org/10.1016/j.mre.2017.03.003
11. G. Mourou, T. Tajima, S.V. Bulanov, **78**, 309 (2009). https://doi.org/10.1103/RevModPhys.78.309
12. A. Di Piazza, C. Müller, C.Z. Hatsagortsyan, C.H. Keitel, **84**, 1177 (2012). https://doi.org/10.1103/RevModPhys.84.1177
13. N.B. Narozhny, A.M. Fedotov, Contemp. Phys. **56**, 249 (2015). https://doi.org/10.1080/00107514.2015.1028768
14. U. Mohideen, M.H. Sher, H.W.K. Tom, G.D. Aumiller, O.R. Wood, R.R. Freeman, J. Bokor, P.H. Bucksbaum, **71**, 509 (1993). https://doi.org/10.1103/PhysRevLett.71.509
15. G.G. Paulus, W. Nicklich, H. Xu, P. Lambropoulos, H. Walther, **72**, 2851 (1994). https://doi.org/10.1103/PhysRevLett.72.2851
16. K. Zrost, A. Rudenko, Th. Ergler, B. Feuerstein, V.L.B. de Jesus, C.D. Schröter, R. Moshammer, J. Ullrich, **39**, S371 (2006). https://doi.org/10.1088/0953-4075/39/13/S10
17. M. Kübel, M. Arbeiter, C. Burger, N.G. Kling, T. Pischke, R. Moshammer, T. Fennel, M.F. Kling, B. Bergues, **51**, 134007 (2018). https://doi.org/10.1088/1361-6455/aac584
18. D.N. Fittinghoff, P.R. Bolton, B. Chang, K.C. Kulander, **69**, 2642 (1992). https://doi.org/10.1103/PhysRevLett.69.2642
19. A. Becker, F.H.M. Faisal, **84**, 3546 (2000). https://doi.org/10.1103/PhysRevLett.84.3546
20. R. Kopold, W. Becker, H. Rottke, W. Sandner, **85**, 3781 (2000). https://doi.org/10.1103/PhysRevLett.85.3781
21. W. Becker, X.J. Liu, P.J. Ho, J.H. Eberly, **84**, 1011 (2012). https://doi.org/10.1103/RevModPhys.84.1011
22. H.G. Hetzheim, C.H. Keitel, **102**, 083003 (2009). https://doi.org/10.1103/PhysRevLett.102.083003
23. H. Bauke, H.G. Hetzheim, G.R. Mocken, M. Ruf, C.H. Keitel, **83**, 063414 (2011). https://doi.org/10.1103/PhysRevA.83.063414
24. E.A. Chowdhury, C.P.J. Barty, B.C. Walker, **63**, 042712 (2001). https://doi.org/10.1103/PhysRevA.63.042712
25. E.A. Chowdhury, B.C. Walker, J. Opt. Soc. Am. B **20**, 109 (2003). https://doi.org/10.1364/JOSAB.20.000109
26. K. Yamakawa, Y. Akahane, Y. Fukuda, M. Aoyama, N. Inoue, H. Ueda, **68**, 065403 (2003). https://doi.org/10.1103/PhysRevA.68.065403
27. K. Yamakawa, Y. Akahane, Y. Fukuda, M. Aoyama, J. Ma, N. Inoue, H. Ueda, H. Kiriyama, **50**, 2515 (2003). https://doi.org/10.1080/09500340308233581
28. A.M. Perelomov, V.S. Popov, M.V. Terentev, Zh. Eksp. Teor. Fiz. **50**, 1393 (1966) (Sov. Phys. JETP **23**, 924 (1966)). http://www.jetp.ac.ru/cgi-bin/dn/e_023_05_0924.pdf
29. M.V. Ammosov, N.B. Delone, V.P. Krainov, Zh. Eksp. Teor. Fiz. **91**, 2008 (1986) (Sov. Phys. JETP **64**, 1191 (1986)). http://www.jetp.ac.ru/cgi-bin/dn/e_064_06_1191.pdf
30. V.S. Popov, Usp. Fiz. Nauk **147**, 921 (2004) (Phys. Usp. **47**, 855 (Engl. transl.) (2004)). https://doi.org/10.3367/UFNr.0174.200409a.0921

31. S.V. Popruzhenko, **47**, 204001 (2014). https://doi.org/10.1088/0953-4075/47/20/204001
32. B.M. Karnakov, V.D. Mur, S.V. Popruzhenko, V.S. Popov, Usp. Fiz. Nauk **185**, 3 (2015) (Phys. Usp. **58**, 3 (2015)). https://doi.org/10.3367/UFNe.0185.201501b.0003
33. Y. Akahane, J. Ma, Y. Fukuda, M. Aoyoma, H. Kiriyama, J.V. Sheldakova, A.V. Kudryashov, K. Yamakawa, Rev. Sci. Instrum. **77**, 023102 (2006). https://doi.org/10.1063/1.2166669
34. M.F. Ciappina, S.V. Popruzhenko, S.V. Bulanov, T. Ditmire, G. Korn, S. Weber, **99** 043405 (2019). https://doi.org/10.1103/PhysRevA.99.043405
35. J. Fletcher, I.R. Cowling, **6**, 258 (1973). https://doi.org/10.1088/0022-3700/6/9/010
36. L.P. Pitaevskii, E.M. Lifshitz, *Physical Kinetics* (Elsevier Science, Amsterdam, 2012)
37. L.V. Keldysh, **47**, 1945 (1964) (Sov. Phys. JETP **20**, 1307 (1965))' http://www.jetp.ac.ru/cgi-bin/dn/e_020_05_1307.pdf
38. D.R. Hartree, Math. Proc. Camb. Philos. Soc. **24**, 89 (1928). https://doi.org/10.1017/S0305004100011919
39. E.G. Gelfer, A.M. Fedotov, S. Weber, Plasma Phys. Control. Fusion **60**, 064005 (2018). https://doi.org/10.1088/1361-6587/aabb12
40. I.Yu. Kostyukov, A.A. Golovanov, **98**, 043407 (2018). https://doi.org/10.1103/PhysRevA.98.043407
41. J. Strohaber, A.A. Kolomenskii, H.A. Schuessler, J. Appl. Phys. **118**, 083107 (2015). https://doi.org/10.1063/1.4929654
42. E.B. Saloman, J. Phys. Chem. Ref. Data **39**, 033101 (2010). https://doi.org/10.1063/1.333766
43. W.H. Press, S.A. Teulkolsky, W.T. Vetterling, B.P. Flannery, *Numerical Recipes in Fortran 90*, vol. 2 (Cambridge University Press, Cambridge, UK, 1996)
44. E.B. Saloman, J. Phys. Chem. Ref. Data **36**, 215 (2007). https://doi.org/10.1063/1.2227036
45. E.B. Saloman, J. Phys. Chem. Ref. Data **33**, 765 (2004). https://doi.org/10.1063/1.1649348

Index

K. Yamanouchi and D. Charalambidis (eds.), *Progress in Ultrafast Intense Laser Science XV*, Topics in Applied Physics 136,
https://doi.org/10.1007/978-3-030-47098-2

Printed by Printforce, the Netherlands